ISBN 978-0-266-82295-0
PIBN 10894685

PHARMACOPŒIA

OF THE

MASSACHUSETTS MEDICAL

SOCIETY.

BOSTON:

PUBLISHED BY E. & J. LARKIN, No. 47, CORNHILL.

GREENOUGH AND STEBBINS, PRINTERS.

1808.

AT a meeting of the Counsellors of the MASSACHUSETTS MEDICAL SOCIETY, held on the 5th day of June, 1807, the Committee, appointed for that purpose, presented the manuscript of a Pharmacopœia, prepared conformably to a vote of the Counsellors, passed on the 3d day of October, 1805; whereon it was voted,

That the said Pharmacopœia be printed for the use of the Society; and that the Committee, who formed it, be appointed to superintend the execution of the printing.

....

THIS work, entitled THE PHARMACOPOEIA OF THE MASSACHUSETTS MEDICAL SOCIETY, *is printed from the manuscript, agreeably to the vote of the Counsellors.*

Boston, December 17, 1807.

James Jackson, } *Committee for the*
John C. Warren, } *Pharmacopœia.*

PREFACE.

THE MASSACHUSETTS MEDICAL SOCIETY present to the public their Pharmacopœia, in conformity with a sense of duty, and the practice of similar bodies of men in Europe. As this is the first work of the kind, which has been published in the United States, they consider it proper to explain the necessity and intention of the work, and to add some particulars respecting its plan and mode of execution.

It is the intention of a Pharmacopœia to point out those articles, whose properties entitle them to be employed for the cure of diseases, with the best modes of preparing them ; and to supply the preparations and compositions of them with titles or names, by which they may be known or referred to, without constantly repeating a description of their ingredients.

Such a work is mutually convenient to the physician and apothecary. As it is the business of the physician to prescribe, and of the apothecary to prepare medicines, the physicians as a body ought to point out those articles of medicine, which they shall ordinarily employ, and the standard preparations of them.

The necessity of a work of this nature has been long noticed, especially in our large towns. In them, the professions of physician and apothecary are most distinct ; and between those, whose relation to each other is so important, a perfect understanding should exist. As this cannot be established between them as individuals, it is necessary that there should be uniformity, both in the pharmaceutical preparations and language. By the want of such uniformity, much inconvenience, and even very serious consequences have been produced.

The work now offered will, it is presumed, effectually remedy such evils. The Medical Society indeed is not empowered to require of apothecaries a compliance with the directions of this Pharmacopœia ; nor does such power seem requisite. It has a sufficient substitute in the apothecary's regard to his own interest, and to his duty to the public.

Books multiply so fast, that it has become necessary to preface every new work with the reasons, almost indeed with an apology, for its appearance. The preceding remarks, furnish in some measure, an apology for this work. Yet it may be thought, that a new book was not necessary for the purposes, which have been stated, since foreign works of this nature are in the hands of every physician and apothecary.

This consideration merits attention, and the Society have endeavoured to give to it due weight. Respecting the medical erudition of those learned colleges, by whom pharmaceutical works have been issued in Europe, the Society very naturally looked to them for a work, which should supply their own wants. They even conceived, that a Pharmacopœia of equal merit, originating from themselves, would be less convenient to them, than one, which should be generally received, and of which the language should be commonly adopted in those nations, with whom we maintain the greatest literary intercourse. It was then a question, what work best answered this description ; and they might have hesitated in their choice among the British Pharmacopœias, had not the Edinburgh college of physicians recently republished theirs upon a plan truly scientific. Yet this, like other European Pharmacopœias, being in the latin language, is not adapted to this country, where the apothecaries are not necessarily instructed in that language; and in some other respects, is not perfectly suited to the modes of practice received among us.

The Society then resolved to adopt the Pharmacopœia of the Edinburgh college as the basis of their own; but to permit such omissions, alterations, and additions as, upon minute examination, should be found necessary. It was not desirable however, to give to this an appearance of originality; on the contrary, trifling considerations have not induced any variation from that excellent work.

As latin names for the articles of the Materia Medica, and for the preparations and compositions, are most commouly employed, these are given, with their English translations; while the directions are in the English language only. The convenience of this arrangement compensates for the partial neglect of system.

Our omissions, alterations, and additions, have not been made without laborious investigations, and such experiments as seemed most important. The following were the general heads of our inquiries.

First, respecting the virtues of each article on the list of the Materia Medica, in the Edinburgh Pharmacopœia.

Second, respecting articles, admitted into other Pharmacopœias, or employed in this country, which are not found in the Edinburgh Pharmacopœia.

Third, respecting the merit of the preparations and compositions in the Edinburgh Pharmacopœia, compared with those, which are similar in other pharmaceutical works.

Fourth, respecting the merit of such preparations and compositions, as are not admitted into the Edinburgh Pharmacopœia, but are found, either in similar works, or in use in our own country.

In pursuing these inquiries, the individuals, to whom the task was allotted, have not been governed by their own sentiments alone; but have constantly had respect to the general opinion of their professional brethren, so far as they could ascertain this by common practice. On the one side, it was obviously desirable, that the book should not be swelled to a great size; and on the other, that it should contain every article of the Materia Medica in use, with sufficient variety of preparations and compositions. Among the substances employed in this country, which are not found in European Pharmacopœias, they have admitted but a small part of those, which have come to their knowledge. It is much to be regretted, that the history of most of them is very imperfect; since such articles only, as have pretensions to an established reputation, can be entitled to admission into a work of this sort. In this instance, a natural partiality has called for an extension of this rule to its utmost length.

On the important subject of pharmaceutical nomenclature, there is much difficulty in determining the best course to be pursued; much more, in pursuing such a course, as to give satisfaction to all. We have followed the Edinburgh college in admitting systematic names, for both the natural and artificial substances, employed in medicine.

It must not be understood, that in adopting the modern language of botany and chemistry, we have consulted the whims and opinions of every pretender. In this, as in former ages, men are creating confusion by creating names. The object of the reforms in scientific language is to obviate this evil, and to establish nomenclature upon solid grounds. The language, we have employed, is that, which has been sanctioned and adopted by the most profoundly scientific men of the age. If chemists have not attained unquestionable knowledge respecting every object of their research, yet this is least of all the case with respect to most of the articles employed in medicine. Shall we refuse to avail ourselves of their successful labours, because they have not arrived at perfection in every branch of their science?

That there are great advantages in the systematic nomenclatures of the botanists and chemists, so far as they are founded on fixed principles, will not be disputed. These advantages are such, as must eventually secure their reception.

There are some cases, in which a regard to convenience has led us to leave the names of well known substances unaltered. Many of those names, which have been established for ages, and much employed, are suffered to remain in use; while the scientific names, corresponding with them, are pointed out in their proper places. The same plan has been followed in those instances, where the scientific names would be too long, or do not point out the variety of the article to be employed with sufficient precision. Thus, it is permitted to call by the name of *opium*, the inspissated juice of the capsules of white poppy; and by the name of *chalk*, the soft carbonate of lime.

The importance of the objects, for which this work is designed, will not be disputed. In the present age, and especially in this country, the articles most employed in medicine are not denied to possess power, whatever doubts may be entertained of their usefulness. Of such articles our knowledge cannot be too perfect, nor can our preparations and compositions be made with too much accuracy. So many are the sources of fallacy in these respects, that all directions are futile, unless a compliance with them is accompanied by great personal accuracy and circumspection in the apothecary. This work is offered as a manual; but it should be handled by those, who are well versed in chemistry, well acquainted with the characters of medicinal substances, and familiarly accustomed to pharmaceutical operations.

Impressed by a sense of duty to their fellow-citizens, the Massachusetts Medical Society may presume that they have in some measure discharged their obligations, if the utility of this work be proportionate to their estimation of its importance. How successful it may be in producing the effects, for which it is designed, must be uncertain. The Society repose with confidence on the hope, that the work will be supported by the well instructed part of the physiciaus and apothecaries; at least, if it be calculated to substitute for obsolete terms and varying forms of preparation, intelligible prescriptions and uniform preparations of medicine. They cannot therefore hesitate to solicit the aid of all scientific men in effecting a revolution, so very desirable for the correct practice of medicine; a revolutiou, which concerns the reputation and success of every medical practitioner, and the health and safety of every individual.

OF WEIGHTS.

AS there frequently arise errors of no small importance from the promiscuous use of weights and measures, it is proper, that the quantities of substances, whether fluid or solid, be determined by weight. Yet it may suffice to measure wine, water, and aqueous liquors in most instances; provided that for this purpose vessels be employed, of glass, where the nature of the substance requires it, whose capacities and divisions accurately correspond with the divisions, or multiples, of the medical pound. The kind of weights, we employ, is that commonly called troy weight, which is divided in the following manner.

A pound	is equal to	twelve ounces.
An ounce	is equal to	eight drachms.
A drachm	is equal to	three scruples.
A scruple	is equal to	twenty grains.

TABLE OF CONTENTS.

PART I.

PART II.

TABLES.

PART I.

MATERIA MEDICA;

OR,

A CATALOGUE OF SIMPLE AND SOME PREPARED MEDICINES, SUCH AS ARE KEPT *IN* THE SHOP OF THE APOTHECARY, BUT NOT PREPARED BY HIM.

MATERIA MEDICA.

ACIDUM acetosum.
Acetous acid.

Acidum sulphuricum.
Sulphuric acid.
The specific gravity of this article, at the temperature of 60° *on Fahrenheit's thermometer, should be to that of distilled water as* 1850 *to* 1000.

Aconitum neomontanum.
Monkshood. *The part used is the herb.*

Acoroides resinifera.
Botany bay gum tree. *The gum resin.*

Acorus calamus.
Sweet flag. *The root.*

Æsculus hippocastanum.
Horse chesnut. *The seed and bark.*

Alcohol.
Alcohol.
The spirit distilled from wine, or other fermented liquors, perfectly free from any unpleasant smell, and of which the specific gravity is to that of water as 835 *to* 1000.

Alcohol dilutum.
Diluted alcohol.
Alcohol, mixed with an equal quantity of water, being somewhat weaker than proof spirit ; its specific gravity is to that of water as 935 *to* 1000.

Allium sativum.
Garlic. *The root.*

Aloe perfoliata. *The gum resin.*
Aloes. *a.* Aloe hepatica. *Hepatic aloes.*
b. Aloe socotorina. *Socotorine aloes.*

Althæa officinalis.

Marsh mallow. *The root and leaves.*

Ammoniacum.

Ammoniacum. *A gum resin.*

Amomum zingiber.

Ginger. *The root, and the candied root brought from India.*

Amomum zedoaria.

Long zedoary. *The root.*

Amomum repens.

Lesser cardamom. *The seeds.*

Amygdalus communis.

The almond tree. *The kernel of the fruit, such as is called sweet almond.*

Amyris gileadensis.

Balsam of gilead tree. *The liquid resin, called balsam of gilead.*

Anethum graveolens.
Dill. *The seeds.*

Anethum fœniculum.
Sweet fennel. *The root and seeds.*

Angelica archangelica.
Garden angelica. *The root, leaves and seeds.*

Angustura.
Augustura. *The bark.*

Anthemis nobilis.
Chamomile. *The flowers.*

Anthemis pyrethrum.
Pellitory of Spain. *The root.*

Apium petroselinum.
Common parsley. *The root.*

*Aqua.
Water.

* By aqua, or water, is always intended pure water.

Arbutus uva ursi.
Bearberry. *The leaves.*

Arctium lappa.
Burdock. *The root and leaves.*

Argentum.
Silver.

Aristolochia serpentaria,
Virginian snakeroot. *The root.*

Arnica montana.
German leopard's bane. *The flower and root.*

Artemisia abrotanum.
Southern wood. *The leaves.*

Artemisia santonica.
Wormseed. *The tops and seeds.*

Artemisia absinthium.
Common wormwood. *The leaves and flowering heads.*

Arum triphyllum.
Indian turnip. *The root.*

Asarum europæum.
Asarabacca. *The leaves.*

Asclepias decumbens.
Pleurisy root. *The root.*

Astragalus tragacantha.
Goatsthorn. *The gum, called gum tragacanth.*

Atropa belladonna.
Deadly nightshade. *The leaves.*

Avena sativa.
Oats. *The seeds.*

Bitumen petroleum.
Rock oil.

Bubon galbanum.
Loveaged leaved bubon. *The gum resin, called galbanum.*

Calx.

Lime, recently burnt.

Cancer pagurus.

The black clawed crab. *The claws, called crab's eyes.*

Canella alba.

Canella alba. *The bark.*

Capsicum annuum.

Cockspur pepper. *The fruit.*

Carbo ligni.

Charcoal of wood.

Carbonas barytæ.

Carbonate of baryta.

Carbonas calcis.

Carbonate of lime.

1. *Soft carbonate of lime, called chalk.*
2. *Indurated carbonate of lime, called marble.*

Carbonas potassæ impurus.

Impure carbonate of potass, called pearl ashes.

Carbonas sodæ impurus.

Impure carbonate of soda.

Carbonas zinci impurus.

Impure carbonate of zinc.

Carum carui.

Caraway. *The seeds.*

Cassia fistula.

Cassia tree. *The fruit.*

Cassia senna.

Senna. *The leaves.*

Castor fiber.

The beaver. *The substance collected in the follicles, near the anus, called castor.*

Centaurea benedicta.

Blessed thistle. *The herb.*

Cera.
Wax.
a. Flava. *Yellow.*
. Alba. *White.*

Cervus elaphus.
The stag, or hart. *The horns.*

Chenopodium anthelminticum.
Jerusalem oak. *The herb and seeds.*

Chironia centaurium.
Smaller centaury. *The flowering heads.*

Cinchona officinalis.
Officinal cinchona. *The bark.*
a. Communis. *The common.*
b. Flava. *The yellow.*
c. Rubra. *The red.*

Cinchona caribæa.
Cinchona of the Caribbean islands. *The bark.*

Citrus aurantium.
Seville orange. *The juice of the fruit and its external rind.*

Citrus medica.
Lemon tree. *The fruit, the rind of the fruit and its volatile oil.*

Coccus cacti.
Cochineal.

Cochlearia officinalis.
Garden scurvy grass. *The plant.*

Cochlearia armoracia.
Horse radish. *The leaves and root.*

Colchicum autumnale.
Meadow saffron. *The root.*

Colomba.
Colomba. *The root.*

Conium maculatum.
Hemlock. *The leaves and seeds.*

Convolvulus scammonia.
Scammony. *The gum resin.*

Convolvulus jalapa.
Jalap. *The root.*

Copaifera officinalis.
Copaiva tree. *The liquid resin, called balsam of copaiva.*

Coriandrum sativum.
Coriander. *The seeds.*

Cornus florida.
Common dogwood. *The flowers, fruit and bark.*

Cornus sericea.
Red willow. *The bark.*

Crocus sativus.
Common saffron. *The summit of the pistils, called saffron.*

Croton eleutheria.
Cascarilla. *The bark.*

Cucumis colocynthis.
Coloquintida. *The medullary part of the fruit.*

Cuminum cyminum.
Cummin. *The seeds.*

Cuprum.
Copper.

Curcuma longa.
Turmeric. *The root.*

Daphnè mezereum.
Mezereon, or spurge laurel. *The bark of the root.*

Datura stramonium.
Thorn apple. *The leaves and seeds.*

Daucus carota.
Wild carrot. *The seeds.*

Delphinium staphisagria.
Stavesacre. *The seeds.*

Digitalis purpurea.
Common fox-glove. *The leaves.*

Dolichos pruriens.
Cowhage. *The stiff hairs which cover the pods.*

Dorstenia contrajerva.
Contrayerva. *The root.*

Eugenia caryophyllata.
The clove tree. *The flower bud, and its volatile oil.*

Ferrum.
Iron.

Ferri oxidum nigrum.
The black oxide of iron, such as are the scales of iron, formed at the foot of the blacksmith's anvil.

Ferula assa fœtida.
Assa fœtida. *The gum resin.*

Ficus carica.
The fig tree. *The fruit.*

Fraxinus ornus.
Manna ash. *The concrete juice, called manna.*

Gambogia.
Gamboge. *A gum resin.*

Gentiana lutea.
Gentian. *The root.*

Geoffræa inermis.
Cabbage bark tree. *The bark.*

Glycyrrhiza glabra.
Liquorice. *The root and extract.*

Gratiola officinalis.
Hedge hyssop. *The herb.*

Guajacum officinale.
Officinal guaiacum. *The wood and resin.*

Hæmatoxylum campechianum.
Logwood tree. *The wood.*

Helleborus niger.
Black hellebore. *The root.*

Helleborus fœtidus.
Bears foot. *The leaves.*

Hordeum distichon.
Barley. *The seed stripped of its husk, called pearl barley.*

Humulus lupulus.
The common hop. *The flowers.*

Hydrargyrus.
Quicksilver.

Hyoscyamus niger.
Black henbane. *The herb and seeds.*

Hyssopus officinalis.
Hyssop. *The herb.*

Inula helenium.
Elecampane. *The root.*

Ipecacuanha.
Ipecacuan. *A root, either of the cephaelis ipecacuanha, or psychotria emetica.*

Iris pseudacorus.
Water flag. *The root.*

Juglans cinerea.
Butternut. *The unripe fruit, and the inner bark.*

Juniperus communis.
Juniper. *The berries and leaves.*

Juniperus lycia.
Olibanum. *The gum resin, called olibanum.*

Juniperus sabina.
Savine. *The leaves.*

Kalmia latifolia.
Broad leaved laurel. *The leaves.*

Kino.
Kino. *A gum resin.*

Lactuca virosa.
Wild lettuce. *The leaves.*

Lactuca sativa.
Common garden lettuce. *The herb.*

Laurus cinnamomum.
The cinnamon tree. *The bark and its volatile oil.*

Laurus cassia.

The cassia tree. *The bark and flower buds gathered before they open.*

Laurus camphora.

Camphor tree. *The Camphor.*

Laurus nobilis.

Bay tree. *The leaves, berries, and fixed oil of the berries.*

Laurus sassafras.

Sassafras. *The wood, root, and its bark.*

Lavandula spica.

Lavender. *The flowering spikes.*

Leontodon taraxacum.

Dandelion. *The root and leaves.*

Lichen islandicus.

Iceland moss. *The herb.*

Linum usitatissimum.

Common flax. *The seeds and their fixed oil.*

Lobelia syphilitica.
Lobelia. *The root.*

Lytta vittata.
Potatoe fly.

Malva sylvestris.
Common mallow. *The leaves and flowers.*

Maranta arundinacea.
Indian arrow root. *The root.*

Marrubium vulgare.
White horehound. *The leaves.*

Melaleuca leucadendron.
The cajeput tree. *The volatile oil.*

Melia azedarach.
Poison berry tree. *The fruit and root.*

Melissa officinalis.
Balm. *The leaves.*

Meloe vesicatorius.
Cantharis.

Mentha viridis.
Spear mint. *The herb.*

Mentha piperita.
Pepper mint. *The herb.*

Mentha pulegium.
Penny royal. *The herb and flower.*

Menyanthes trifoliata.
Marsh trefoil. *The leaves.*

Mimosa catechu.
Catechu. *The extract of the wood, called extract of catechu.*

Mimosa nilotica.
Egyptian mimosa. *The gum, called gum arabic.*

Momordica elaterium.
Wild cucumber. *The fresh fruit when almost ripe.*

Moschus moschiferus.
The musk deer. *The substance contained in a follicle situated near the navel, called musk.*

Murias ammoniæ.
Muriate of ammonia.

Murias sodæ.
Muriate of soda.

Myristica moschata.
The nutmeg tree. *The kernel of the fruit, called nutmeg; its involucre, called mace; its fixed oil, called oil of mace; and its volatile oil.*

Myroxylon peruiferum.
Sweet smelling balsam tree. *The balsam, called peruvian balsam.*

Myrrha.
Myrrh. *A gum resin.*

Myrtus pimenta.
Pimento tree. *The fruit, called jamaica pepper.*

Nicotiana tabacum.
Tobacco. *The leaves.*

Nitras potassæ.
Nitrate of potass.

Olea europæa.
The olive tree. *The fixed oil of the fruit, called olive oil.*

Origanum majorana.
Sweet marjorum. *The herb.*

Ovis aries.
The sheep. *The fat, called mutton suet.*

Oxalis acetosella.
Wood sorrel. *The leaves.*

Oxidum arsenici.
Oxide of arsenic.

Oxidum plumbi album.
White oxide of lead.

Oxidum plumbi rubrum.
Red oxide of lead.

Oxidum plumbi semivitreum.
Semi-vitrified oxide of lead.

Oxidum zinci impurum.
Impure oxide of zinc.

Papaver somniferum.
White poppy. *The capsules and their inspissated juice, called opium.*

Physeter macrocephalus.
Spermaceti whale. *The matter found within the cranium, called spermaceti.*

Phytolacca decandra.
American nightshade. *The leaves and berries.*

Pimpinella anisum.
Anise. *The seeds.*

Pinus abies.
Common spruce fir. *The resin which concretes spontaneously, called burgundy pitch.*

Pinus balsamea.

Balsam fir. — *The liquid resin, called balsam of Canada.*

Pinus larix.

The larch. — *The liquid resin, called venice turpentine, and volatile oil, called oil of turpentine.*

Pinus sylvestris.

Scotch fir. — *The empyreumatic resin, called tar; and the liquid resin, called common turpentine.*

Piper nigrum.

Black pepper. — *The fruit.*

Piper longum.

Long pepper. — *The fruit.*

Pistacia lentiscus.

Mastich tree. — *The resin, called mastich.*

Plumbum.

Lead.

Podophyllum peltatum.
May apple. *The root.*

Polygala senega.
Seneka. *The root.*

Polygonum bistorta.
Great bistort. *The root.*

Polypodium filix mas.
Male fern. *The root.*

Prunus domestica.
The french prune tree. *The fruit, called french prunes.*

Prunus Virginiana.
Wild cherry tree. *The bark, both of the tree and root.*

Pterocarpus santalinus.
Red saunders tree. *The wood.*

Pterocarpus draco.
The resin, called dragons' blood.

Quassia simaruba.
Mountain damson. *The bark.*

Quassia excelsa.
Quassia. *The wood, bark, and root.*

Quercus robur.
Oak. *The bark.*

Quercus cerris.
Oriental oak. *The nest of the cynips quercifolii, called gall nut.*

Resina pini.
Resin of pine. *A resin, which is procured from pines of various species, deprived of its volatile oil.*

Rhamnus catharticus.
Purging buckthorn. *The juice of the berries.*

Rheum palmatum.
Palmated rhubarb. *The root.*

Rhododendron crysanthum.
Yellow flowered rhododendron. *The leaves.*

Rhus toxicodendron.
Poison oak. *The leaves.*

Ricinus communis.
Palma christi. *The seeds and their fixed oil, called castor oil.*

Rosa gallica.
Red rose. *The petals.*

Rosa damascena.
Damask rose. *The petals.*

Rosa canina.
Dog rose. *The fruit, called hips.*

Rosmarinus officinalis.
Rosemary. *The flowering tops.*

Rubia tinctorum.
Madder. *The root.*

Ruta graveolens.
Rue. *The herb.*

Saccharum officinarum.

Sugar cane. *Sugar.*

a. *brown, or impure.*

b. *double refined, or most pure.*

Sagapenum.

Sagapenum. *A gum resin.*

Salvia officinalis.

Sage. *The leaves.*

Sambucus nigra.

Common elder. *The inner bark, flowers and berries.*

Sanguinaria canadensis.

Blood root. *The seed and root.*

Sapo.

Soap, prepared with oil of olives and soda, called castile soap.

Scilla maritima.

Squill. *The root.*

Sinapis alba.
White mustard. *The seed.*

Smilax sarsaparilla.
Sarsaparilla. *The root.*

Solanum dulcamara.
Bitter sweet. *The twigs.*

Spigelia marilandica.
Carolina pink. *The root.*

Spongia officinalis.
Sponge.

Stannum.
Tin. *The filings and powder.*

Styrax officinalis.
Officinal storax. *The balsam.*

Styrax benzoin.
Benjamin tree. *The balsam, called benzoin.*

Sub acetis cupri.
Sub acetite of copper.

Sub-boras sodæ.

Sub borate of soda.

Succinum.

Amber.

Super sulphas aluminæ et potassæ.

Super sulphate of alumina and potass.

Sulphas barytæ.

Sulphate of baryta.

Sulphas cupri.

Sulphate of copper.

Sulphas magnesiæ.

Sulphate of magnesia.

Sulphur sublimatum.

Sublimed sulphur.

Sulphuretum antimonii.

Sulphuret of antimony.

Super-tartris potassæ.

Super-tartrite of potass.

Super-tartris potassæ impurus.
Impure super-tartrite of potass.

Sus scrofa.
The hog. *The fat, called hog's lard.*

Tamarindus indica.
Tamarind tree. *The preserved fruit, called tamarinds.*

Tanacetum vulgare.
Tansy. *The flowers and leaves.*

Toluifera balsamum.
Balsam of tolu tree. *The balsam, called balsam of tolu.*

Tormentilla erecta.
Septfoil. *The root.*

Triticum æstivum.
Wheat. *The flour and starch prepared from the seeds.*

Tussilago farfara.
Coltsfoot. *The leaves and flowers.*

Ulmus fulva.
Slippery elm. *The inner bark.*

Valeriana officinalis.
Wild valerian. *The root.*

Veratrum album.
White hellebore. *The root.*

Viola odorata.
March violet. *The recent flower.*

Vitis vinifera.
The vine. *The dried fruit, called raisin, and the fermented juice of the fruit, called spanish white wine.*

Xanthoxylum clava herculis.
Toothach tree. *The bark and seed vessels.*

Zincum.
Zinc.

PART II.

PREPARATIONS AND COMPOSITIONS.

PREPARATIONS AND COMPOSITIONS.

CHAP. I.

SULPHUR.

SULPHUR SUBLIMATUM LOTUM.

WASHED SUBLIMED SULPHUR.

Take of

Sublimed sulphur, one pound,

Water, four pounds.

Boil the sulphur for a little while in the water, then pour off this water, and wash away all the acid by affusions of cold water; lastly, dry the sulphur.

CHAP. II.

ACIDS, ALKALIES, EARTHS, AND THEIR COMPOUNDS.

ACIDUM SULPHURICUM DILUTUM.

DILUTED SULPHURIC ACID.

Take of

Sulphuric acid, one ounce,

Water, seven ounces.

Mix them gradually.

ACIDUM NITROSUM.

NITROUS ACID.

Take of

Nitrate of potass in coarse powder, or bruised, two pounds,

Sulphuric acid, sixteen ounces.

Having put the nitrate of potass into a glass retort, pour upon it the sulphuric acid, and distil it in a sand bath, with a heat gradually increased, until the iron pot begins to be red hot.

The specific gravity of this acid is to that of distilled water as 1550 to 1000.

ACIDUM NITROSUM DILUTUM.

DILUTED NITROUS ACID.

Take of

Nitrous acid,

Water, equal weights.

Mix them, taking care to avoid the noxious vapours.

ACIDUM NITRICUM.

NITRIC ACID.

Take of

Nitrous acid, any quantity.

Pour it into a retort, and having adapted a receiver, apply a very gentle heat, until the reddest portion shall have passed over, and the acid which remains in the retort shall have become nitric acid.

ACIDUM MURIATICUM.

MURIATIC ACID.

Take of

Muriate of soda, two pounds,

Sulphuric acid, sixteen ounces,

Water, one pound.

Let the muriate of soda be kept at a red heat for some time in an iron vessel, and after it has cooled, put it into a retort; then pour upon the muriate of soda, the acid mixed with the water and allowed to cool. Lastly, distil in a sand bath, with a moderate fire, as long as any acid is produced.

The specific gravity of this acid is to that of distilled water, as 1170 to 1000.

ACIDUM ACETOSUM DESTILLATUM.

DISTILLED ACETOUS ACID.

Let eight pounds of acetous acid be distilled in glass vessels with a gentle heat. The two first pounds which come over, being too watery, are to be set aside; the next four pounds will be the distilled acetous acid. The remainder furnishes a still stronger acid, but too much burnt by the fire.

ACIDUM ACETOSUM FORTE.

STRONG ACETOUS ACID.

Take of

Sulphate of iron dried, one pound,
Acetite of lead, ten ounces.

Having rubbed them together, put them into a retort, and distil in a sand bath with a moderate heat, as long as any acid comes over.

ACIDUM BENZOICUM.

BENZOIC ACID.

Take of

Benzoin, twenty-four ounces,
Carbonate of soda, eight ounces,
Water, sixteen pounds.

Triturate the benzoin with the carbonate, then boil in the water for half an hour, with constant agitation, and strain. Repeat the decoction, with other six pounds of water, and strain. Mix these decoctions, and evaporate, until two pounds remain. Filter anew, and drop into the fluid, as long as it produces any precipitation,

Diluted sulphuric acid.

Dissolve the precipitated benzoic acid in boiling water; strain the boiling solution through linen, and set it aside to crystallize. Wash the crystals with cold water, dry and preserve them.

OLEUM SUCCINI ET ACIDUM SUCCINICUM.

OIL OF AMBER AND SUCCINIC *ACID.*

Take of

Amber reduced to powder, and of pure sand, equal weights.

Mix them, and put them into a glass retort, of

which the mixture may fill one half ; then adapt a large receiver, and distil in a sand bath, with a fire gradually increased. At first a watery liquor will come over, with some yellow oil ; then a yellow oil with an acid salt, and lastly, a reddish and black coloured oil. Pour the liquor out of the receiver, and separate the oil from the water. Press the salt collected from the neck of the retort and sides of the receiver, between folds of blotting paper, to free it from the oil adhering to it ; then purify it by solution in warm water and crystallization.

AQUA ACIDI CARBONICI.

WATER OF CARBONIC ACID.

Take of

Water, six pounds ; place this in the middle part of a Nooth's apparatus, and expose it to a stream of carbonic acid gas arising from

Carbonate of lime in powder,

Sulphuric acid, each three ounces,

Water, three pounds, gradually and cautiously mixed.

If a larger quantity of the liquor be required, the apparatus of Dr. Woulfe is preferable.

In this and similar preparations, where carbonic acid gas is combined with liquids, the liquor is better in proportion to the coldness of the air, and to the pressure to which it is subjected. It should be preserved in glass vessels well closed, and should not be expofed to any high temperature.

AQUA POTASSÆ.

SOLUTION OF POTASS.

Take of

Lime recently burnt, eight ounces,

Carbonate of potass, six ounces.

Throw the lime into an iron or earthen vessel, with twenty-eight ounces of warm water. After the ebullition is finished, instantly add the salt; and having thoroughly mixed them, cover the vessel till they cool. When the mixture has cooled, agitate it well, and pour it into a glass funnel, whose throat must be stopt up with a piece of clean rag. Let the upper mouth of the funnel be covered, while the tube of it is inserted into another glass vessel, so that the solution of potass may gradually drop through the rag into the lower vessel. When it first gives over dropping, pour into the funnel some ounces of water; but cautiously, so that the water may swim above the matter. The water of potass will again begin to drop, and the affusion of water is to be repeated in the same manner, until three pounds have dropped, which will happen in the space of two or three days ; then by agitation mix the superior and inferior parts of the liquor together, and put it up in a well stopt phial.

POTASSA.

POTASS.

Take of

Solution of potass, any quantity.

Evaporate it in a covered very clean iron vessel, till on the ebullition ceasing, the saline matter flows gently like oil, which happens before the vessel becomes red. Then pour it out on a smooth iron plate ; let it be divided into small pieces before it hardens, and immediately placed in a well stopt phial.

POTASSA CUM CALCE.

POTASS WITH LIME.

Take of

Solution of potass, any quantity.

Evaporate this in a covered iron vessel till one third remains ; then mix with it as much new slacked lime as will bring it to the consistence of a pretty solid pap, which is to be kept in a vessel closely stopt.

CARBONAS POTASSÆ.

CARBONATE OF POTASS.

Let impure carbonate of potass, put into a crucible, be brought to a low red heat, that the oily impurities, if there be any, may be consumed; then triturate it with an equal weight of water, and mix them thoroughly by agitation. Filtrate the liquor through paper into a very clean iron pot, and boil to dryness, stirring the salt towards the end of the process, to prevent its sticking to the vessel.

CARBONAS POTASSÆ PURISSIMUS.

PURE CARBONATE OF POTASS.

Take of

Impure super-tartrite of potass, any quantity.

Burn it to a black mass, by placing it among live coals, either wrapped up in moist bibulous paper, or contained in a crucible. Having reduced this mass to powder, expose it in an open crucible to the action of a moderate fire, till it become white, or at least of an ash grey colour, taking care that it do not melt. Then dissolve it in warm water; strain the liquor through a linen cloth, and evaporate it in a clean iron vessel, diligently stirring it towards the end of the process with an iron spatula, to prevent it from sticking to the bottom of the vessel. A very white salt will remain, which is to be left a little longer on the fire, till the bottom of the vessel becomes almost red. Lastly, when the salt is grown cold, keep it in glass vessels well stopt.

AQUA CARBONATIS POTASSÆ.

SOLUTION OF CARBONATE OF POTASS.

Take of

Carbonate of potass, any quantity.

Set it in a moist place till it deliquesce, and then strain it.

AQUA SUPER-CARBONATIS POTASSÆ.

SOLUTION OF SUPER-CARBONATE OF POTASS.

Take of

Water, ten pounds,

Pure carbonate of potass, one ounce.

Dissolve and expose the solution to a stream of

carbonic acid gas, in the same manner as directed for the water of carbonic acid.

ACETIS POTASSÆ.

ACETITE OF POTASS.

Take of

Pure carbonate of potass, one pound.

Boil it with a very gentle heat, in four or five times its weight of distilled acetous acid; add more acid at different times, till on the watery part of the preceding quantity being nearly dissipated by evaporation, the new addition of acid ceases to raise any effervescence; which will happen, when about twenty pounds of the distilled acetous acid have been consumed. It is then to be slowly dried. The impure salt remaining, is to be melted with a gentle heat, for a short time; and afterwards dissolved in water, and filtered through paper. If the liquifaction has been properly performed, the filtered liquor will be limpid; but if otherwise, of a brown colour. Afterwards evaporate this liquor with a very gentle heat in a very shallow glass vessel, occasionally stirring the salt as it becomes dry, that its moisture may be sooner dissipated. Lastly, the acetite of potass ought to be kept in a vessel very closely stopt, to prevent it from deliquescing.

SULPHAS POTASSÆ.

SULPHATE OF POTASS.

Take of

Sulphuric acid diluted, with six times its weight of water, any quantity.

Put it into a capacious glass vessel, and gradually drop into it, of pure carbonate of potass, dissolved in six times its weight of water, as much as is sufficient thoroughly to saturate the acid. The effervescence being finished, strain the liquor through paper; and after due evaporation set it aside to crystallize.

Sulphate of potass may be also conveniently prepared from the residuum of the distillation of nitrous acid, by dissolving it in warm water, and saturating it with carbonate of potass.

SULPHAS POTASSÆ CUM SULPHURE.

SULPHATE OF POTASS WITH SULPHUR.

Take of

Nitrate of potass in powder,

Sublimed sulphur, of each equal parts.

Mix them well together, and inject the mixture, by little and little at a time, into a red hot crucible; the deflagration being over, let the salt cool, after which it is to be put up in a glass vessel well stopped.

SULPHURETUM POTASSÆ.

SULPHURET OF POTASS.

Take of

Carbonate of potass,

Sublimed sulphur, each eight ounces.

Having ground them well together, put them into a large coated crucible; and having fitted a cover to it, and applied live coals cautiously around it, bring them at length to a state of fusion.

Having broken the crucible as soon as it has grown cold, take out the sulphuret, and keep it in a well closed phial.

TARTRIS POTASSÆ.

TARTRITE OF POTASS.

Take of

Carbonate of potass, one pound,

Super-tartrite of potass, three pounds, or as much as may be sufficient,

Boiling water, fifteen pounds.

To the carbonate of potass dissolved in the water, gradually add the super-tartrite of potass in fine powder, as long as it raises any effervescence, which generally ceases before three times the weight of the carbonate of potass has been added ; then strain the cooled liquor through paper, and after due evaporation set it aside to crystallize.

CARBONAS SODÆ.

CARBONATE OF SODA.

Take of

Impure carbonate of soda any quantity.

Bruise it ; then boil in water till all the salt be dissolved. Strain the solution through paper, and evaporate it in an iron vessel, so that after it has cooled, the salt may crystallize.

AQUA SUPER-CARBONATIS SODÆ.

SOLUTION OF SUPER-CARBONATE OF SODA.

Take of

Water, ten pounds,

Carbonate of soda, two ounces.

Dissolve and expose the solution to a stream of carbonic acid gas, in the same manner as directed for the water of carbonic acid.

PHOSPHAS SODÆ.

PHOSPHATE OF SODA.

Take of

Bones burnt to whiteness, and powdered, ten pounds,

Sulphuric acid, six pounds,

Water, nine pounds.

Mix the powder with the sulphuric acid in an earthen vessel; then add the water and mix again. Then place the vessel in a vapour bath, and digest for three days; after which dilute the mass with nine pounds more of boiling water, and strain the liquor through a strong linen cloth, pouring over it boiling water, in small quantities at a time, until the whole acid be washed out.

Set by the strained liquor, that the impurities may subside, decant the clear solution, and evaporate it to nine pounds. To this liquor, poured from the impurities, and heated in an earthen vessel, add carbonate of soda, dissolved in warm water, until the effervescence cease. Filter the neutralized liquor, and set it aside to crystallize. To the liquor that remains after the crystals are taken out, add a little carbonate of soda, if necessary, so as to saturate exactly the phosphoric acid, and dispose the liquor by evaporation to form crystals as long as these can be produced. Lastly, the crystals are to be kept in a well closed vessel.

SULPHAS SODÆ.

SULPHATE OF SODA.

Dissolve the acidulous salt which remains after the distillation of muriatic acid, in water ; and having mixed chalk with it to remove the superfluous acid, set it aside until the sediment subside ; then decant the liquor, strain it through paper, and evaporate it so that it may crystallize.

TARTRIS POTASSÆ ET SODÆ.

TARTRITE OF POTASS AND SODA.

It is prepared from the carbonate of soda and super-tartrite of potass, in the same manner as the tartrite of potass.

AQUA AMMONIÆ.

WATER OF AMMONIA.

Take of

Muriate of ammonia, one pound,
Lime, fresh burnt, one pound and an half,
Distilled water, one pound,
Water, nine ounces.

Pour the water on the powdered lime contained in an iron or earthen vessel, which is then to be covered up till the lime falls to powder. Then mix the muriate previously ground into very fine powder, thoroughly with the lime, by triturating them together in a mortar, and immediately put the mixture into a retort of bottle glass. Put the retort in a sand bath, and connect with it a Woulfe's apparatus. In the first and smallest bottle, furnished with a tube of safety, put two ounces of the distilled water, and in the second the rest of the distilled water.

The fire is now to be kindled, and gradually increased, until the bottom of the sand pot becomes red. Mix the fluid contained in each of the bottles, and preserve it in small phials accurately closed.

ALCOHOL AMMONIATUM.

AMMONIATED ALCOHOL.

Take of

Alcohol, thirty-two ounces,
Lime, fresh burnt, twelve ounces,
Muriate of ammonia, eight ounces,
Water, eight ounces.

From these ingredients, ammoniated alcohol is

prepared, in exactly the same manner, as the water of ammonia.

CARBONAS AMMONIÆ.

CARBONATE OF AMMONIA.

Take of

Muriate of ammonia, one pound,

Pure soft carbonate of lime dried, two pounds.

Having triturated them separately, mix them thoroughly, and sublime from a retort into a refrigerated receiver.

AQUA CARBONATIS AMMONIÆ.

SOLUTION OF CARBONATE OF AMMONIA.

Take of

Muriate of ammonia,

Carbonate of potass, each sixteen ounces,

Water, two pounds.

Having mixed the salts and put them into a glass retort, pour the water upon them, and distil to dryness in a sand bath, gradually increasing the heat.

AQUA ACETITIS AMMONIÆ.

WATER OF ACETITE OF AMMONIA.

Take of

Carbonate of ammonia in powder, any quantity.

Pour upon it as much distilled acetous acid as may be sufficient to saturate the ammonia exactly.

HYDRO-SULPHURETUM AMMONIÆ.

HYDRO SULPHURET OF AMMONIA.

Take of

Water of ammonia, four ounces, subject it in a chemical apparatus to a stream of the gas, which arises from

Sulphuret of iron, four ounces,

Muriatic acid, eight ounces, previously diluted with two pounds and a half of water.

Sulphuret of iron is conveniently prepared for this purpose, from

Purified filings of iron, three parts,

Sublimed sulphur, one part.

Mixed and exposed to a moderate degree of heat in a covered crucible, until they unite into a mass.

MURIAS BARYTÆ.

MURIATE OF BARYTA.

Take of

Carbonate of baryta,
Muriatic acid, each one part,
Water, three parts.

Add the carbonate, broken into little bits, to the water and acid, previously mixed. After the effervescence has ceased, digest for an hour, strain the liquor, and set it aside to crystallize. Repeat the evaporation as long as any crystals are formed.

If the carbonate of baryta cannot be procured, the muriate may be prepared in the following manner from the sulphate.

Take of

Sulphate of baryta, two pounds,
Charcoal of wood in powder, four ounces.

Roast the sulphate with fire, that it may be more easily reduced to a very fine powder, with which the charcoal is to be intimately mixed. Put the mixture into a crucible, and having fitted it with a cover, heat it with a strong fire for six hours. Then triturate the matter well, and throw it into six pounds of water, in an earthen or glass vessel, and mix them by agitation, preventing as much as possible the access of the air.

Let the veseel stand in a vapour bath until the part not dissolved shall subside, then pour off the liquor ; on the undissolved part pour four pounds more of boiling water, which, after agitation and deposition, are to be added to the former liquor. Into the liquor while still warm, or if it shall have cooled, again heated, drop muriatic acid as long as it excites any effervescence. Then strain it and evaporate it so as to crystallize.

SOLUTIO MURIATIS BARYTÆ.

SOLUTION OF MURIATE OF BARYTA.

Take of

Muriate of baryta, one part,

Distilled water, three parts.

Dissolve.

AQUA CALCIS.

LIME WATER.

Take of

Lime recently burnt, half a pound.

Put it into an earthen vessel, and sprinkle on it four ounces of water, keeping the vessel shut, while the lime grows hot, and falls into powder. Then pour on it twelve pounds of water, and mix the lime thoroughly with the water, by agitation. After the lime has subsided, repeat the agitation; and let this be done about ten times, always keeping the vessel shut, that the free access of the air may be prevented. Lastly, let the water be filtered through paper, placed in a funnel, with glass rods interposed between them, that the water may pass as quickly as possible.

It must be kept in very close bottles.

CARBONAS CALCIS PRÆPARATUS.

PREPARED CARBONATE OF LIME.

Carbonate of lime, whether the variety, commonly called chalk, or that called crab's eyes and crab's stones, after having been triturated to powder in an iron mortar, and levigated on a porphyry stone, with a little water, is to be put into a large vessel, and water to be poured upon it; which, after agitating the vessel repeatedly, is to be again poured off, while loaded with fine powder. On allowing the water to settle, a subtile powder will subside, which is to be dried.

The coarse powder which the water could not

suspend, may be levigated again, and treated in the same manner.

SOLUTIO MURIATIS CALCIS.

SOLUTION OF MURIATE OF LIME.

Take of

Hard carbonate of lime, that is white marble, broken into pieces, nine ounces,

Muriatic acid, sixteen ounces,

Water, eight ounces.

Mix the acid with the water, and gradually add the pieces of carbonate of lime. When the effervescence has ceased, digest them for an hour; pour off the liquor and evaporate it to dryness. Dissolve the residuum in its weight and a half of water; and, lastly, filter the solution.

PHOSPHAS CALCIS IMPURUS.

IMPURE PHOSPHATE OF LIME.

Burn pieces of hartshorn till they become perfectly white; then reduce them to a very fine powder.

CARBONAS MAGNESIÆ.

CARBONATE OF MAGNESIA.

Take of

Sulphate of magnesia,

Carbonate of potass, equal weights.

Dissolve them separately in double their quantity of warm water, and let the liquors be strained or otherwise freed from the feces; then mix them and instantly add eight times their quantity of warm water. Let the liquor boil for a little on the fire, stirring it at the same time; then let it rest till the heat be somewhat diminished; after which strain it through linen; the carbonate of magnesia will remain upon the cloth, and it is to be washed with pure water till it become altogether void of saline taste.

MAGNESIA.

MAGNESIA.

Let carbonate of magnesia, put into a crucible, be kept in a red heat for two hours; then put it up in close stopt glass vessels.

SUPER-SULPHAS ALUMINÆ ET POTASSÆ EXSICCATUS.

DRIED SUPER-SULPHATE OF ALUMINA AND POTASS.

Melt super-sulphate of alumina and potass in an earthen or iron vessel, and keep it over the fire until it eease to boil.

CHAP. III.

METALLINE PREPARATIONS.

ANTIMONY.

SULPHURETUM ANTIMONII PRÆPARATUM.

PREPARED SULPHURET OF ANTIMONY.

Sulphuret of antimony is prepared in the same manner as carbonate of lime, (p. 58.)

OXIDUM ANTIMONII CUM SULPHURE, PER NITRATEM POTASSÆ.

OXIDE OF ANTIMONY, WITH SULPHUR, BY NITRATE OF POTASS.

Take of

Sulphuret of antimony,

Nitrate of potass, equal weights.

After they are separately powdered and well mixed, let them be injected into a red hot crucible; when the deflagration is over, separate the reddish metallic matter from the whitish crust; powder it and edulcorate it by repeated washings with hot water, till the water come off insipid.

OXIDUM ANTIMONII, CUM SULPHURE, VITRIFICATUM.

VITRIFIED OXIDE OF ANTIMONY WITH SULPHUR.

Strew sulphuret of antimony beat into a coarse powder like sand, upon a shallow unglazed earthen vessel, and apply a gentle fire underneath, that the sulphuret may be heated slowly; keeping it at the same time continually stirring, to prevent it from running into lumps. White vapours of sulphureous smell will arise from it. When they cease with the degree of heat first applied, increase the fire a little, so that the vapours may again arise; go on in the same manner, till the powder, when brought to a red heat, exhales no more vapours. Melt this powder in a crucible with an intense heat, till it assumes the appearance of melted glass; then pour it out on a heated brass plate.

OXIDUM ANTIMONII VITRIFICATUM, CUM CERA.

VITRIFIED OXIDE OF ANTIMONY WITH WAX.

Take of

Yellow wax, one part,
Vitrified oxide of antimony with sulphur, eight parts.

Melt the wax in an iron vessel, and throw into it

the powdered oxide ; roast the mixture over a gentle fire for a quarter of an hour, continually stirring it ; then pour it out, and when cold, grind it into powder.

SULPHURETUM ANTIMONII PRÆCIPITATUM.

PRECIPITATED SULPHURET OF ANTIMONY.

Take of

Solution of potass, four pounds,
Water, three pounds,
Prepared sulphuret of antimony, two pounds.

Boil them in a covered iron pot, over a slow fire for three hours, adding more water if necessary, and frequently stirring the mixture with an iron spatula ; strain the liquor while warm through a double cloth, and add to it when filtered, as much diluted sulphuric acid as is necessary to precipitate the sulphuret, which must be well washed with warm water.

OXIDUM ANTIMONII CUM PHOSPHATE CALCIS.

OXIDE OF ANTIMONY WITH PHOSPHATE OF LIME.

Take of

Sulphuret of antimony in coarse powder,
Shavings of hartshorn, equal weights.

Mix, and put them into a wide red-hot iron pot, and stir the mixture constantly, until it be burnt into a matter of a grey colour, which is then to be removed from the fire, ground into powder, and put into a coated crucible. Lute to this crucible another inverted over it, and perforated in the bottom with a small hole, and apply the fire, which is to be raised gradually to a white heat, and kept in that increased state for two hours. Lastly, grind the matter, when cold, into a very fine powder.

MURIAS ANTIMONII.

MURIATE OF ANTIMONY.

Take of

Oxide of antimony with sulphur, by ui-trate of potass,

Sulphüric acid, each one pound,

Dried muriate of soda, two pounds.

Pour the sulphuric acid into a retort, gradually adding the muriate of soda and oxide of antimony previously mixed. Then perform the distillation in a sand bath. Expose the distilled matter for several days to the air, that it may deliquesce, and then pour the liquid part from the feces.

TARTRIS ANTIMONII.

TARTRITE OF ANTIMONY.

Take of

Oxide of antimony with sulphur, by nitrate of potass, three parts,

Super-tartrite of potass, four parts,

Distilled water, thirty-two parts.

Boil in a glass vessel for a quarter of an hour, strain through paper, and set aside the strained liquor to crystallize.

SILVER.

NITRAS ARGENTI.

NITRATE OF SILVER.

Take of

Purest silver, flatted into plates, and cut in pieces, four ounces,

Diluted nitrous acid, eight ounces,

Distilled water, four ounces.

Dissolve the silver in a phial with a gentle heat, and evaporate the solution to dryness. Then put the mass into a large crucible, and place it on the fire, which should at first be gentle, and afterward increased by degrees, till the mass flows like oil ;

then pour it into iron pipes, previously heated and anointed with tallow. Lastly, let it be kept in a glass vessel well closed.

COPPER.

AMMONIARETUM CUPRI.

AMMONIARET OF COPPER.

Take of

Purest sulphate of copper, two parts,
Carbonate of ammonia, three parts.

Rub them carefully together in a glass mortar, until after the effervescence has entirely ceased; they unite into a violet coloured mass, which must be wrapped up in blotting paper, and first dried on a chalk stone, and afterwards by a gentle heat. The product must be kept in a glass phial well closed.

SOLUTIO SULPHATIS CUPRI COMPOSITA.

COMPOUND SOLUTION OF SULPHATE OF COPPER.

Take of

Sulphate of copper,
Super-sulphate of alumina and potass, each three ounces,
Water, two pounds,
Diluted sulphuric acid, an ounce and a half.

Boil the sulphates in the water to dissolve them, and then add the acid to the liquor filtered through paper.

IRON.

FERRI LIMATURA PURIFICATA.

PURIFIED FILINGS OF IRON.

Let a piece of pure iron be filed with a clean file, then place a sieve over the filings, and apply a magnet, so that the filings may be attracted upwards through the sieve.

OXIDUM FERRI NIGRUM PURIFICATUM.

PURIFIED BLACK OXIDE OF IRON.

Let the scales of the oxide of iron, which are to be found at the foot of the blacksmith's anvil, be purified by the application of a magnet. For the magnet will attract only the smaller and purer scales, and will leave those which are larger and less pure.

CARBONAS FERRI.

CARBONATE OF IRON.

Moisten purified filings of iron frequently with water, that they may be converted into rust, which is to be ground into an impalpable powder.

CARBONAS FERRI PRÆCIPITATUS.

PRECIPITATED CARBONATE OF IRON.

Take of

Sulphate of iron, four ounces,
Carbonate of soda, five ounces,
Water, ten pounds.

Dissolve the sulphate in the water, and add the carbonate of soda, previously dissolved, in a sufficient quantity of water, and mix them thoroughly. Wash the carbonate of iron, which is precipitated, with warm water, and afterwards dry it.

AQUA SUPER-CARBONATIS FERRI.

SOLUTION OF THE SUPER-CARBONATE OF IRON.

It is prepared in the same manner as the water of carbonic acid, by suspending in the water half an ounce of iron wire.

SULPHAS FERRI.

SULPHATE OF IRON.

Take of

Purified filings of iron, six ounces,
Sulphuric acid, eight ounces,
Water, two pounds and a half.

Mix them, and after the effervescence ceases, digest the mixture for some time upon warm sand; then strain the liquor through paper, and after due evaporation, set it at rest to crystallize.

SULPHAS FERRI EXSICCATUS.

DRIED SULPHATE OF IRON.

Take of

Sulphate of iron, any quantity.

Expose it to the action of a moderate heat in an unglazed earthen vessel, until it become white and perfectly dry. The heat applied here must not be so great as to decompose the sulphate of iron, but only to deprive it of its water of crystallization.

OXIDUM FERRI RUBRUM.

RED OXIDE OF IRON.

Expose dried sulphate of iron to an intense heat, until it is converted into a very red matter.

MURIAS AMMONIÆ ET FERRI.

MURIATE OF AMMONIA AND IRON.

Take of

Red oxide of iron, washed and again dried,
Muriate of ammonia, equal weights.

Mix them thoroughly and sublime.

QUICKSILVER.

HYDRARGYRUS PURIFICATUS.

PURIFIED QUICKSILVER.

Take of

Quicksilver, four parts,
Purified filings of iron, one part.

Rub them together, and distil from an iron vessel.

ACETIS HYDRARGYRI.

ACETITE OF QUICKSILVER.

Take of

Purified quicksilver, three ounces,
Diluted nitrous acid, four ounces and a half, or a little more than may be required for dissolving the quicksilver,
Acetite of potass, three ounces,
Boiling water, eight pounds.

Mix the quicksilver with the diluted nitrous acid, and after the effervescence has ceased, digest if necessary with a gentle heat, until the quicksilver be entirely dissolved. Then dissolve the acetite of potass in the boiling water, and immediately to this solution, still hot, add the former, and mix them by agitation. Then set the mixture aside to crystallize. Place the crystals in a funnel, and wash them with cold distilled water; and, lastly, dry them with as gentle a heat as possible.

In preparing the acetite of quicksilver, it is important that all the vessels and tunnels employed, be of glass.

MURIAS HYDRARGYRI.

MURIATE OF QUICKSILVER.

Take of

Purified quicksilver, two pounds,
Sulphuric acid, two pounds and a half,
Dried muriate of soda, four pounds.

Boil the quicksilver with the sulphuric acid in a glass vessel, placed in a sand bath, until the matter be dried. Mix the matter when cold, in a glass vessel, with the muriate of soda, then sublime in a glass cucurbit, with a heat gradually increased. Lastly, separate the sublimed matter from the scoriæ.

SUB-MURIAS HYDRARGYRI.

SUB-*MURIATE OF QUICKSILVER.*

Take of

Muriate of quicksilver, ground to powder in a glass mortar, four ounces,

Purified quicksilver, three ounces.

Rub them together in a glass mortar, with a little water, to prevent the acrid powder from rising, until the quicksilver be extinguished ; and having put the powder, after being dried, into an oblong phial, of which it fills one third, sublime from warm sand. After the sublimation is finished, having broken the phial, throw away both the red matter found near the bottom of the phial, and the white matter near its neck, and sublime the rest of the mass. Grind this into a very minute powder, which is, lastly, to be washed with boiling distilled water.

SUB-MURIAS HYDRARGYRI PRÆCIPITATUS.

PRECIPITATED SUB-*MURIATE OF QUICKSILVER.*

Take of

Diluted nitrous acid,

Purified quicksilver, each eight ounces,

Muriate of soda, four ounces and a half,

Boiling water, eight pounds.

Mix the quicksilver with the diluted nitrous acid, and towards the end of the effervescence digest with a gentle heat, frequently shaking the vessel in the mean time. But it is necessary to add more quicksilver to the acid than it is capable of dissolving, that a perfectly saturated solution may be obtained.

Dissolve at the same time, the muriate of soda in the boiling water, and into this solution pour the other, while still hot, and mix them quickly by agitation. Pour off the saline liquor after the precipitate has subsided, and wash the sub-muriate of quicksilver by repeated affusions of boiling water, which is to be poured off each time after the deposition of the sub-muriate, until the water come off tasteless.

SUB-MURIAS HYDRARGYRI ET AMMONIÆ.

SUB-*MURIATE OF QUICKSILVER AND AMMONIA*.

Take of

Muriate of quicksilver,
Muriate of ammonia,
Solution of carbonate of potass, each half a pound.

Dissolve first the muriate of ammonia, afterwards

the muriate of quicksilver, in distilled water, and add to these the solution of carbonate of potass. Filtrate and wash the powder until it become insipid.

OXIDUM HYDRARGYRI CINEREUM.

ASH-COLOURED OXIDE OF QUICKSILVER.

Take of

Purified quicksilver, four parts,
Diluted nitrous acid, five parts,
Distilled water, fifteen parts,
Solution of carbonate of ammonia, a sufficient quantity.

Dissolve the quicksilver in the nitrous acid ; then gradually add the distilled water, and pour into the mixture as much water of the carbonate of ammonia as shall be sufficient to precipitate the whole of the oxide of quicksilver ; which is then to be washed with pure water and dried.

OXIDUM HYDRARGYRI RUBRUM, PER ACIDUM NITRICUM.

RED OXIDE OF QUICKSILVER, BY NITRIC ACID.

Take of

Purified quicksilver, one pound,
Diluted nitrous acid, sixteen ounces.

Dissolve the quicksilver, and evaporate the solution, with a gentle heat, to a dry white mass; which, after being ground into powder, is to be put into a glass cucurbit, and to have a thick glass plate laid upon its surface. Then, having adapted a capital, and placed the vessel in a sand bath, apply a gradually increased heat, until the matter be converted into very red scales.

SUB-SULPHAS HYDRARGYRI FLAVUS.

YELLOW SUB-SULPHATE OF QUICKSILVER.

Take of

Purified quicksilver, four ounces,
Sulphuric acid, six ounces.

Put them into a glass cucurbit, and boil them in a sand bath to dryness. Throw into boiling water the white matter, which is left in the bottom, after having reduced it to powder. A yellow powder will immediately be produced, which must be frequently washed with warm water.

SULPHURETUM HYDRARGYRI NIGRUM.

BLACK SULPHURET OF QUICKSILVER.

Take of

Purified quicksilver,
Sublimed sulphur, each equal weights.

Grind them together in a glass mortar, with a glass pestle, till the globules of quicksilver totally disappear.

It is also prepared with twice the quantity of quicksilver.

SULPHURETUM HYDRARGYRI RUBRUM.

RED SULPHURET OF QUICKSILVER.

Take of

Purified quicksilver, forty ounces,
Sublimed sulphur, eight ounces.

Mix the quicksilver with the melted sulphur, and if the mixture take fire, extinguish it by covering the vessel ; afterwards reduce the mass to powder, and sublime it.

LEAD.

ACETIS PLUMBI.

ACETITE OF LEAD.

Take of

White oxide of lead, any quantity ; put it into a cucurbit, and pour upon it of
Distilled acetous acid, ten times its weight.

Let the mixture stand upon warm sand till the

acid become sweet; when it is to be poured off, and fresh acid added until it cease to become sweet. Then evaporate all the liquor, freed from impurities, in a glass vessel, to the consistence of thin honey, and set it aside in a cold place, that the crystals may be formed, which are to be dried in the shade. The remaining liquor is again to be evaporated, that new crystals may be formed; and the evaporation is to be repeated until no more crystals concrete.

TIN.

PULVIS AMALGAMATIS STANNI.

POWDER OF THE AMALGAM OF TIN.

Take of

Tin, five parts,

Purified quicksilver, two parts,

Prepared carbonate of lime, one part.

Melt the tin, add to it the quicksilver, and rub them together; then add the carbonate of lime, and while the mixture is liquid, rub till the metallic particles disappear. While the mixture cools reduce it to an impalpable powder.

ZINC.

OXIDUM ZINCI.

OXIDE OF ZINC.

Let a large crucible be placed in a furnace filled with live coals, so as to be somewhat inclined towards its mouth; and when the bottom of the crucible is moderately red, throw into it a small piece of zinc, about the weight of a drachm. The zinc soon inflames, and it is at the same time converted into white flakes, which are from time to time to be removed from the surface of the metal with an iron spatula, that the combustion may be more complete; and at last, when the zinc ceases to flame, the oxide of zinc is to be taken out of the crucible. Having put in another piece of zinc, the operation is to be repeated, and may be repeated as often as is necessary. Lastly, the oxide of zinc is to be prepared in the same way as the carbonate of lime.

CARBONAS ZINCI IMPURUS PRÆPARATUS.

PREPARED IMPURE CARBONATE OF ZINC.

The impure carbonate of zinc, after being roasted by those who make brass, is prepared in the same way as carbonate of lime.

OXIDUM ZINCI IMPURUM PRÆPARATUM.

PREPARED IMPURE OXIDE OF ZINC.

This is prepared in the same manner as carbonate of lime.

SULPHAS ZINCI.

SULPHATE OF ZINC.

Take of

Zinc, cut into small pieces, three ounces,
Sulphuric acid, five ounces,
Water, twenty ounces.

Mix them, and when the effervescence is finished, digest the mixture for a little on hot sand; then strain the decanted liquor through paper, and after proper evaporation, set it apart, that it may crystallize.

SOLUTIO ACETITIS ZINCI.

SOLUTION OF ACETITE OF ZINC.

Take of

Sulphate of zinc, a drachm,

Distilled water, ten ounces.

Dissolve.

Take of

Acetite of lead, four scruples,

Distilled water, ten ounces.

Dissolve.

Mix the solutions ; let them stand at rest a little, and then filter the liquor.

CHAP. IV.

ETHER, AND ETHERIAL SPIRITS.

ÆTHER SULPHURICUS.

SULPHURIC ETHER.

Take of

Sulphuric acid,

Alcohol, each thirty-two ounces.

Pour the alcohol into a glass retort fit for sustaining a sudden heat, and add to it the acid in an uninterrupted stream.

Mix them by degrees, shaking them moderately and frequently: this done, instantly distil from sand previously heated for the purpose, into a receiver kept cool with water or snow. But the heat is to be so managed, that the liquor shall boil as soon as possible, and continue to boil till sixteen ounces are drawn off; then let the retort be removed from the sand.

To the distilled liquor add two drachms of potass; then distil from a very high retort, with a very gentle heat, into a cool receiver, until ten ounces have been drawn off. If sixteen ounces of alcohol be poured upon the acid remaining in the retort after the first distillation, and the distillation be repeated, more ether will be obtained, and this may be repeated several times.

ÆTHER SULPHURICUS CUM ALCOHOLE.

SULPHURIC ETHER WITH ALCOHOL.

Take of

Sulphuric ether, one part,

Alcohol, two parts.

Mix them.

SPIRITUS ÆTHERIS NITROSI.

SPIRIT OF NITROUS ETHER.

Take of

Alcohol, three pounds,

Nitrous acid, one pound.

Pour the alcohol into a capacious phial, placed in a vessel full of cold water, and add the acid by degrees, constantly agitating them. Let the phial be slightly covered, and placed for seven days in a cool place; then distil the liquor with the heat of boiling water into a receiver kept cool with water or snow, as long as any spirit comes over.

CHAP. V.

THE DRYING OF HERBS AND FLOWERS.

Herbs and flowers are to be dried by the gentle heat of a stove or common fire, in such quantities at a time, that the process may be finished as quickly as possible ; for in this manner their properties are best preserved ; the test of which is the perfect preservation of their natural colour.

The leaves of hemlock (conium maculatum) and of other plants containing a subtile volatile matter, must be immediately pounded, after being dried, and afterwards kept in glass phials, well corked.

SCILLA MARITIMA EXSICCATA.

DRIED SQUILL.

Cut the root of the squill, after having removed its external coat, transversely into thin slices, and dry it by a gentle heat.

The sign of its being properly dried is that although rendered friable, it retains its bitterness and acrimony.

CHAP. VI.

EXPRESSED AND INSPISSATED JUICES.

SUCCUS COCHLEARIÆ COMPOSITUS.

COMPOUND JUICE OF SCURVY GRASS.

Take of

Juice of garden scurvy grass,
——— water cresses, expressed from fresh gathered herbs,
——— seville oranges, of each two pounds,
Spirit of nutmegs, half a pound.

Mix them, and let them stand till the feces have subsided; then pour off the clear liquor.

SUCCUS SPISSATUS ACONITI NEOMONTANI.

INSPISSATED JUICE OF MONKSHOOD.

Bruise the fresh leaves of monkshood; and including them in a hempen bag, compress them strongly till they yield their juice; which is to be evaporated in flat vessels heated with boiling water, saturated with muriate of soda, and immediately reduced to the consistence of thick honey. After the mass has become cold, let it be put up in glazed earthen vessels, and moistened with alcohol.

In the same manner are prepared from their leaves,

Succi Spissati	*The inspissated juices of*
Atropæ belladonnæ,	Deadly nightshade.
Conii maculati,	Hemlock.
Daturæ stramonii,	Thorn apple.
Hyosciami nigri,	Black henbane.
Lactucæ virosæ,	Wild lettuce.
Lactucæ sativæ.	Common garden lettuce.

SUCCUS SPISSATUS SAMBUCI NIGRÆ.

INSPISSATED JUICE OF ELDER BERRIES.

Take of

Juice of the ripe berries of common elder, five pounds,

Purest sugar, one pound.

Evaporate with a gentle heat to the consistence of pretty thick honey.

SUCCUS SPISSATUS MOMORDICÆ ELATERII.

INSPISSATED JUICE OF THE WILD CUCUMBER.

Cut into slices ripe wild cucumbers, and pass the juice, very lightly expressed, through a very fine hair sieve ; then boil it a little and set it by for some hours, until the thicker part has subsided.

Pour off the thinner part swimming at the top, and separate the rest by filtration. Cover the thicker part which remains after the filtration, with a linen cloth, and dry it with a gentle heat.

PULPARUM EXTRACTIO.

THE EXTRACTION OF PULPS.

Boil unripe pulpy fruits, and ripe ones, if they be dry, in a small quantity of water until they become soft ; then press out the pulb through a hair sieve, and afterwards boil it down to the consistence of honey in an earthen vessel, in a water bath, stirring the matter continually, that it may not burn.

The pulp of cassia fistularis is in like manner to be boiled out from the bruised pod, and reduced afterwards to a proper consistence, by evaporating the water.

The pulps of fruits that are both ripe and fresh, are to be pressed out through the sieve, without any previous boiling.

CHAP. VII.

FIXED OILS, AND OILY PREPARATIONS.

OLEUM AMYGDALI COMMUNIS.

OIL OF ALMONDS.

Take of

Fresh sweet almonds, any quantity.

After having bruised them in a stone mortar, put them into a hempen bag, and express the oil without heat.

In the same manner is to be expressed from its seeds,

OLEUM LINI USITATISSIMI.

OIL OF LINSEED.

OLEUM AMMONIATUM.

AMMONIATED OIL.

Take of

Olive oil, two ounces,

Water of ammonia, two drachms.

Mix them together.

The above preparation may be made with three or four times the quantity of water of ammonia.

OLEUM SULPHURATUM.

SULPHURATED OIL.

Take of

Olive oil, eight ounces,

Sublimed sulphur, one ounce.

Boil them together in a large iron pot, stirring them continually till they unite.

OLEUM CAMPHORATUM.

CAMPHORATED OIL.

Take of

Olive oil, two ounces,

Camphor, half an ounce.

Mix them so that the camphor may be dissolved.

CHAP. VIII.

DISTILLED WATERS, AND SPIRITS.

AQUA DESTILLATA.

DISTILLED WATER.

Let water be distilled in very clean vessels, until about two thirds of it have come over.

AQUA CITRI AURANTII.

ORANGE PEEL WATER.

Take of

Fresh orange peel, two pounds.

Pour upon it as much water as shall be sufficient to prevent any empyreuma, after ten pounds have been drawn off by distillation.

After due maceration, distil ten pounds.

The same quantity of water is to be distilled in the same manner, from

Six pounds of the recent petals of the Damask rose, tò prepare *Aqua rosæ damascenæ.*

Damask rose water.

Three pounds of

Peppermint in flower......*Aqua menthæ piperitæ.*

Peppermint water.

Three pounds of

Pennyroyal in flower.......*Aqua menthæ pulegii.*

Pennyroyal water.

One pound and a half of

Spearmint in flower.........*Aqua menthæ viridis.*

Spearmint water.

One pound of

The bark of cinnamon.....*Aqua lauri cinnamomi.*

Cinnamon water.

One pound of

The bark of cassia.........*Aqua lauri cassiæ.*

Cassia water.

Half a pound of

The fruit of pimento......*Aqua myrti pimentæ.*

Pimento water.

To each pound of these waters add half an ounce of diluted alcohol.

SPIRITUS CARI CARUI.

SPIRIT *OF CARAWAY.*

Take of

Caraway seeds bruised, half a pound,

Diluted alcohol, nine pounds.

Macerate two days in a close vessel; then pour on as much water as will prevent empyreuma, and draw off by distillation nine pounds.

In the same manner is prepared the same quantity of spirit from

One pound of *Spiritus*

Bark of cinnamon, bruised.....*Lauri cinnamomi.*
Spirit of cinnamon.

One pound and a half of

Peppermint in flower...........*Menthæ piperitæ.*
Spirit of peppermint.

One pound and a half of

Spearmint in flower............*Menthæ viridis.*
Spirit of spearmint.

Two ounces of

Nutmeg, well bruised..........*Myristicæ moschatæ.*
Spirit of nutmeg.

Half a pound of

The fruit of pimento bruised..*Myrti pimentæ.*
Spirit of pimento.

SPIRITUS LAVANDULÆ SPICÆ.

SPIRIT OF LAVENDER.

Take of

Flowering spikes of lavender, fresh gathered, two pounds,

Alcohol, eight pounds.

Draw off by the heat of boiling water, seven pounds.

By these directions, and in the same quantity is prepared,

SPIRITUS RORISMARINI OFFICINALIS.

SPIRIT *OF ROSEMARY.*

SPIRITUS JUNIPERI COMPOSITUS.

COMPOUND SPIRIT OF JUNIPER.

Take of

Juniper berries, well bruised, one pound,

Caraway seeds,

Sweet fennel seeds, each bruised, one ounce and a half,

Diluted alcohol, nine pounds,

Water, sufficient to prevent empyreuma.

Macerate two days, and draw off nine pounds by distillation.

ALCOHOL AMMONIATUM FŒTIDUM.

FETID AMMONIATED ALCOHOL.

Take of

Ammoniated alcohol, eight ounces,

The gum resin of assa fœtida, half an ounce.

Digest in a close vessel twelve hours; then distil off with the heat of boiling water eight ounces.

CHAP. IX.

VOLATILE OILS.

Volatile oils are prepared nearly in the same manner as the distilled waters, except that less water is to be added. Seeds and woody substances are to be previously bruised or rasped.

The oil comes over with the water, and is afterwards to be separated from it, according as it may be lighter than the water, and swim upon its surface, or heavier, and sink to the bottom. Besides, in preparing both distilled waters and oils, it is to be observed, that the goodness of the subject, its texture, the season of the year, and similar causes, must give rise to so many differences, that no certain or general rule can be given to suit accurately each example.

Therefore, many things are omitted, to be varied by the operator according to his judgment, and only the most general precepts are given.

According to these directions are prepared the

Volatile oils of	*Olea volatilia*
Anise seeds. .	*Pimpinellæ anisi.*

Fennel seeds.	*Anethi fæniculi.*
Juniper berries.	*Juniperi communis.*
Pimento.	*Myrti pimentæ.*
Rosemary.	*Rorismarini officinalis.*
Lavender.	*Lavendulæ spicæ.*
Peppermint.	*Menthæ piperitæ.*
Spearmint.	*Menthæ viridis.*
Pennyroyal.	*Menthæ pulegii.*

OLEUM VOLATILE PINI PURISSIMUM.

PURIFIED VOLATILE OIL OF FINE.

Take of

Oil of turpentine, one pound,

Water, four pounds.

Distil as long as any oil comes over.

OLEUM SUCCINI PURISSIMUM.

PURIFIED OIL OF AMBER.

Distil oil of amber in a glass retort, with six times its quantity of water, till two thirds of the water have passed into the receiver; then separate this very pure volatile oil from the water, and keep it for use in well closed vessels.

CHAP. X.

INFUSIONS AND DECOCTIONS.

INFUSUM CINCHONÆ OFFICINALIS.

INFUSION OF CINCHONA.

Take of

Bark of cinchona, in coarse powder, one ounce,

Mucilage of gum arabic, two ounces,

Water, one pound.

Triturate the cinchona with the mucilage, and add the water during the trituration. Macerate for twenty-four hours, and decant the pure liquor.

INFUSUM DIGITALIS PURPUREÆ.

INFUSION OF COMMON FOX-GLOVE.

Take of

Dried leaves of common fox-glove, one drachm,

Boiling water, eight ounces,

Spirit of cinnamon, one ounce.

Macerate for four hours, and filtrate.

INFUSUM GENTIANÆ COMPOSITUM.

COMPOUND INFUSION OF GENTIAN.

Take of

Root of gentian sliced, half an ounce,
Rind of seville oranges dried and bruised, one drachm,
Coriander seeds bruised, half a drachm,
Diluted alcohol, four ounces,
Water, one pound.

First pour on the alcohol, and three hours after add the water ; then macerate without heat for twelve hours and strain.

INFUSUM MIMOSÆ CATECHU.

INFUSION OF CATECHU.

Take of

Extract of catechu in powder, two drachms and a half,
Bark of cinnamon bruised, half a drachm,
Boiling water, seven ounces,
Simple syrup, one ounce.

Macerate the extract and cinnamon in the hot water, in a covered vessel, for two hours ; then strain it and add the syrup.

INFUSUM RHEI PALMATI.

INFUSION OF RHUBARB.

Take of

Root of rhubarb, bruised, half an ounce,
Boiling water, eight ounces,
Spirit of cinnamon, one ounce.

Macerate the rhubarb in a close vessel with the water for twelve hours ; then having added the spirit, strain the liquor.

INFUSUM ROSÆ GALLICÆ.

INFUSION OF ROSES.

Take of

Petals of red roses, dried, two ounces,
Boiling water, five pounds,
Sulphuric acid, one drachm,
Double refined sugar, two ounces.

Macerate the petals with the boiling water in an earthen vessel, which is not glazed with lead, for four hours ; then having poured on the acid, strain the liquor and add the sugar.

INFUSUM TAMARINDI ET SENNÆ.

INFUSION OF TAMARINDS AND SENNA.

Take of

Preserved tamarinds, one ounce,
Leaves of senna, one drachm,
Coriander seeds bruised, half a drachm,
Brown sugar, half an ounce,
Boiling water, eight ounces.

Macerate them for four hours, occasionally agitating them in a close earthen vessel, not glazed with lead, and strain the liquor.

It may also be made with double, triple, &c. the quantity of senna.

DECOCTUM ALTHÆÆ OFFICINALIS.

DECOCTION OF MARSH-MALLOW.

Take of

Root of marsh-mallow, dried and bruised, four ounces,
Raisins stoned, two ounces,
Water, seven pounds.

Boil to five pounds; place apart the strained liquor till the feces have subsided, then pour off the clear liquor.

DECOCTUM CINCHONÆ OFFICINALIS.

DECOCTION OF CINCHONA.

Take of

Bark of cinchona in powder, one ounce,
Water, a pound and a half.

Boil for ten minutes in a covered vessel, and strain the liquor while hot.

DECOCTUM DAPHNES MEZEREI.

DECOCTION OF MEZEREON.

Take of

Bark of the root of mezereon, two drachms,
Root of liquorice bruised, half an ounce,
Water, three pounds.

Boil with a gentle heat to two pounds and strain.

DECOCTUM GUAIACI COMPOSITUM.

COMPOUND DECOCTION OF GUAIACUM.

Take of

Raspings of the wood of guaiacum, three ounces,
Raisins, two ounces,
Root of sassafras sliced,
Root of liquorice bruised, each one ounce,
Water, ten pounds.

Boil the guaiacum and raisins with the water, over a gentle fire, to the consumption of one half; adding towards the end the sassafras and liquorice. Strain the liquor without expression.

DECOCTUM HORDEI DISTICHI.

DECOCTION OF BARLEY.

Take of

Pearl barley, two ounces,

Water, five pounds.

First wash the barley from the mealy matter that adheres to it, with some cold water; then boil it a little with about half a pound of water, to extract the colouring matter. Throw this away, and put the barley thus purified into five pounds of boiling water; which is to be boiled down to one half and strained.

DECOCTUM POLYGALÆ SENEGÆ.

DECOCTION OF SENEKA.

Take of

Root of seneka, one ounce,

Water, two pounds.

Boil to sixteen ounces and strain.

DECOCTUM SMILACIS SARSAPARILLÆ.

DECOCTION OF SARSAPARILLA.

Take of

Root of sarsaparilla sliced, six ounces,

Water, eight pounds.

Macerate for two hours with a heat of about 195° ; then take out the root and bruise it ; return the bruised root to the liquor, and again maccrate it for two hours. Then the liquor being boiled to the measure of four pints, press it out and strain.

CHAP. XI.

EMULSIONS AND MUCILAGES.

EMULSIO AMYGDALI COMMUNIS.

EMULSION OF ALMONDS.

Take of

Sweet almonds blanched, one ounce,

Water, two pounds and a half.

Beat the almonds very thoroughly in a stone mortar, gradually pouring on them the water; then strain off the liquor.

EMULSIO CAMPHORATA.

EMULSION OF CAMPHOR.

Take of

Camphor, one scruple,

Sweet almonds blanched, two drachms,

Double refined sugar, one drachm,

Water, six ounces.

This is to be made in the same manner as the almond emulsion.

EMULSIO AMMONIACI.

EMULSION OF AMMONIACUM.

Take of

Ammoniacum, two drachms,

Water, eight ounces.

Rub the gum resin with the water, gradually poured on, until it become an emulsion.

MISTURA CARBONATIS CALCIS.

MIXTURE OF CARBONATE OF LIME.

Take of

Prepared carbonate of lime, one ounce,

Double refined sugar, half an ounce,

Mucilage of gum arabic, two ounces.

Triturate together, and then gradually add of

Water, two pounds and a half,

Spirit of cinnamon, two ounces.

Mix them.

MUCILAGO AMYLI.

MUCILAGE OF STARCH.

Take of

Starch, half an ounce,

Water, one pound.

Triturate the starch, gradually adding the water; then boil them a little.

MUCILAGO ASTRAGALI TRAGACANTHÆ.

MUCILAGE OF GUM TRAGACANTH.

Take of

Gum tragacanth in powder, one ounce,
Boiling water, eight ounces.

Macerate twenty-four hours ; then triturate them carefully, that the gum may be dissolved, and press the mucilage through linen cloth.

MUCILAGO MIMOSÆ NILOTICÆ.

MUCILAGE OF GUM ARABIC.

Take of

Gum arabic in powder, one part,
Boiling water, two parts.

Digest with frequent agitation, until the gum be dissolved ; then press the mucilage through linen.

CHAP. XII.

SYRUPS.

SYRUPUS SIMPLEX.

SIMPLE SYRUP.

Take of

Double refined sugar, fifteen parts,

Water eight parts.

Let the sugar be dissolved in the water by a gentle heat, and boiled a little, so as to form a syrup.

SYRUPUS ALTHÆÆ OFFICINALIS.

SYRUP OF MARSH-MALLOW.

Take of

Fresh roots of marsh-mallow sliced, one pound,

Water, ten pounds,

Double refined sugar, four pounds.

Boil the water with the roots to the consumption of one half and strain the liquor strongly expressing it. Suffer the strained liquor to rest till the feces have subsided; and to the depurated liquor add the sugar; then boil so as to make a syrup.

SYRUPUS AMOMI ZINGIBERIS.

SYRUP OF GINGER.

Take of

Roots of ginger bruised, three ounces,

Boiling water, four pounds,

Double refined sugar, seven pounds and a half.

Macerate the ginger in the water in a close vessel, for twenty-four hours; then to the liquor strained, add the sugar in powder, so as to make a syrup.

SYRUPUS CITRI AURANTII.

SYRUP OF ORANGE PEEL.

Take of

The fresh outer rind of seville oranges, six ounces,

Boiling water, three pounds,

Double refined sugar, four pounds.

Macerate the rind in the water for twelve hours; then add to the filtrated liquor the sugar in powder, and apply a gentle heat, so as to form a syrup.

SYRUPUS CITRI MEDICÆ.

SYRUP OF LEMONS.

Take of

Juice of lemons, suffered to stand till the feces have subsided, and afterwards strained, three parts,

Double refined sugar, five parts.

Dissolve the sugar in the juice, so as to make a syrup.

SYRUPUS PAPAVERIS SOMNIFERI.

SYRUP OF WHITE POPPIES.

Take of

Capsules of white poppy dried, and freed from the seeds, two pounds,

Boiling water, thirty pounds,

Double refined sugar, four pounds.

Macerate the sliced capsules in the water for twelve hours; next boil till only one third part of the liquor remain; then strain it by expressing it strongly. Boil the strained liquor to the consumption of one half, and strain again. Lastly, add the sugar, and boil a little, so as to form a syrup.

SYRUPUS RHAMNI CATHARTICI.

SYRUP OF BUCKTHORN.

Take of

Juice of ripe buckthorn berries depurated, two parts,

Double refined sugar, one part.

Boil them so as to form a syrup.

SYRUPUS ROSÆ DAMASCENÆ.

SYRUP OF DAMASK ROSES.

Take of

Fresh petals of the damask rose, one pound,

Boiling water, four pounds,

Double refined sugar, three pounds.

Macerate the roses in the water for a night; then to the liquor strained and freed from the dregs, add the sugar; boil them into a syrup.

SYRUPUS ROSÆ GALLICÆ.

SYRUP OF RED ROSES.

Take of

Dried petals of red roses, seven ounces,

Double refined sugar, six pounds,

Boiling water, five pounds.

Macerate the roses in the water for twelve hours, then boil them a little and strain the liquor ; add to it the sugar, and boil them again for a little so as to form a syrup.

SYRUPUS SCILLÆ MARITIMÆ.

SYRUP OF SQUILLS.

Take of

Acetous acid with squills, two pounds,

Double refined sugar in powder, three pounds and a half.

Dissolve the sugar with a gentle heat, so as to form a syrup.

SYRUPUS TOLUIFERÆ BALSAMI.

SYRUP OF BALSAM OF TOLU.

Take of

Simple syrup, two pounds,

Tincture of balsam of tolu, one ounce,

With the syrup recently prepared, and when it has almost grown cold, after it has been removed from the fire, gradually mix the tincture, with constant agitation.

SYRUPUS VIOLÆ ODORATÆ.

SYRUP OF VIOLETS.

Take of

Flowers of fresh violets, one pound,
Boiling water, four pounds,
Double refined sugar, seven pounds and a half.

Macerate the violets in the water for twenty-four hours in a glass or a glazed earthen vessel, close covered; then strain without expression, and to the strained liquor add the sugar, powdered, and make into a syrup.

CHAP. XIII.

MEDICATED VINEGARS.

ACETUM AROMATICUM.

AROMATIC ACETOUS ACID.

Take of

Tops of rosemary dried,
Leaves of sage dried, each four ounces,
Flowers of lavender dried, two ounces,
Cloves in coarse powder, two drachms,
Distilled acetous acid, eight pounds.

Macerate for seven days, express the liquor and filtrate through paper.

ACETUM SCILLÆ MARITIMÆ.

ACETOUS ACID WITH SQUILLS.

Take of

Dried root of squills, two ounces,
Distilled acetous acid, two pounds and a half,
Alcohol, three ounces.

Macerate the squills with the acetous acid for seven days; then press out the liquor, to which add the alcohol; and when the feces have subsided, pour off the clear liquor.

A preparation similar to that formerly kept under the name of *oxymel of squills*, may be made by mixing extemporaneously any quantity of honey with the above preparation.

ACIDUM ACETOSUM CAMPHORATUM.

CAMPHORATED ACETOUS ACID.

Take of

Strong acetous acid, six ounces,

Camphor, half an ounce.

Reduce the camphor to powder, by triturating it with a little alcohol; then add it to the acid, in which it should be dissolved.

CHAP. XIV.

TINCTURES.

TINCTURA ALOES SOCOTORINÆ.

TINCTURE OF SOCOTORINE ALOES.

Take of

Socotorine aloes in powder, half an ounce,
Extract of liquorice, an ounce and a half,
Alcohol, four ounces,
Water, one pound.

Digest for seven days in a closed vessel, with a gentle heat and frequent agitation, and when the feces have subsided, decant the tincture.

These directions are to be observed in preparing all tinctures.

TINCTURA ALOES ET MYRRHÆ.

TINCTURE OF ALOES AND MYRRH.

Take of

Myrrh in powder, two ounces,
Alcohol, one pound and a half,
Water, half a pound.

Mix the alcohol with the water, then add the myrrh; digest for four days; and lastly add of

Socotorine aloes in powder,
Saffron sliced, each two ounces.

Digest again for three days, and pour off the tincture from the sediment.

TINCTURA AMOMI REPENTIS.

TINCTURE OF CARDAMOM.

Take of

Lesser cardamom seeds bruised, four ounces,
Diluted alcohol, two pounds and a half.

Digest for seven days, and filtrate through paper.

TINCTURA ARISTOLOCHIÆ SERPENTARIÆ.

TINCTURE *OF VIRGINIAN SNAKEROOT.*

Take of

Root of Virginian snakeroot bruised, three ounces,
Cochineal in powder, one drachm,
Diluted alcohol, two pounds and a half.

Digest for seven days, and filtrate through paper.

TINCTURA FERULÆ ASSÆ FŒTIDÆ.

TINCTURE OF ASSA FOETIDA.

Take of

Gum resin of assa fœtida, four ounces,

Alcohol, two pounds and a half.

Digest for seven days, and filtrate through paper.

TINCTURA BENZOIN COMPOSITA.

COMPOUND TINCTURE OF BENZOIN.

Take of

Benzoin in powder, three ounces,

Peruvian balsam, two ounces,

Hepatic aloes in powder, half an ounce,

Alcohol, two pounds.

Digest with a gentle heat for seven days, and filtrate through paper.

TINCTURA CAMPHORÆ.

TINCTURE OF CAMPHOR.

Take of

Camphor, one ounce,

Alcohol, one pound.

Mix them together, that the camphor may be dissolved.

It may also be made with a double or triple proportion of camphor.

TINCTURA ELEUTHERIÆ.

TINCTURE OF CASCARILLA.

Take of

Bark of cascarilla in powder, four ounces,
Diluted alcohol, two pounds.

Digest with a gentle heat for eight days and filtrate.

TINCTURA ANGUSTURÆ.

TINCTURE OF ANGUSTURA.

Take of

Bark of angustura in powder, four ounces,
Diluted alcohol, two pounds.

Digest with a gentle heat for eight days and filtrate.

TINCTURA SENNÆ COMPOSITA.

COMPOUND TINCTURE OF SENNA.

Take of

Leaves of senna, three ounces,
Root of jalap bruised, one ounce,
Coriander seeds,
Caraway seeds, each bruised, half an ounce,
Lesser cardamom seeds bruised, two drachms,
Diluted alcohol, three pounds and a half.

Digest for seven days, and to the liquor filtrated through paper add of

Double refined sugar, four ounces.

TINCTURA CASTOREI.

TINCTURE OF CASTOR.

Take of

Russian castor in powder, an ounce and a half,

Alcohol, one pound.

Digest for seven days and filtrate through paper.

TINCTURA CINCHONÆ OFFICINALIS.

TINCTURE OF CINCHONA.

Take of

Red bark of cinchona in powder, six ounces,

Diluted alcohol, two pounds and a half.

Digest for seven days and filtrate through paper.

TINCTURA CINCHONÆ COMPOSITA.

COMPOUND TINCTURE OF CINCHONA.

Take of

Red bark of cinchona in powder, two ounces,

External rind of seville oranges dried, one ounce and a half,

Root of virginian snakeroot bruised, three drachms,

Saffron, one drachm,

Cochineal in powder, two scruples,

Diluted alcohol, twenty ounces.

Digest for fourteen days and filtrate.

TINCTURA COLOMBÆ.

TINCTURE OF *COLOMBA.*

Take of

Colomba root in powder, two ounces and a half,

Diluted alcohol, two pounds and an half.

Digest for seven days and filtrate through paper.

TINCTURA CONVOLVULI JALAPÆ.

TINCTURE OF JALAP.

Take of

Root of jalap in powder, four ounces,

Diluted alcohol, fifteen ounces.

Digest for seven days and filtrate through paper.

TINCTURA CROCI SATIVI.

TINCTURE OF SAFFRON.

Take of

English saffron sliced, one ounce,

Diluted alcohol, fifteen ounces.

Digest for seven days and filtrate through paper.

TINCTURA DIGITALIS PURPUREÆ.

TINCTURE OF COMMON FOX-GLOVE.

Take of

Dried leaves of common fox-glove, one ounce,

Diluted alcohol, eight ounces.

Digest for seven days and filtrate through paper.

TINCTURA GENTIANÆ COMPOSITA.

COMPOUND TINCTURE OF GENTIAN.

Take of

Root of gentian sliced and bruised, two ounces,

Rind of seville oranges dried and bruised, one ounce,

Bark of canella alba bruised, half an ounce,

Powder of cochineal, half a drachm,

Diluted alcohol, two pounds and a half.

Digest for seven days and filtrate through paper.

TINCTURA GUAJACI OFFICINALIS.

TINCTURE *OF GUAIACUM.*

Take of

Gum resin of officinal guaiacum in powder, one pound,

Alcohol, two pounds and a half.

Digest for seven days and filtrate through paper.

TINCTURA HELLEBORI NIGRI.

TINCTURE *OF BLACK HELLEBORE.*

Take of

Root of black hellebore bruised, four ounces,

Cochineal in powder, half a drachm,

Diluted alcohol, two pounds and a half.

Digest seven days and filtrate through paper.

TINCTURA HYOSCIAMI NIGRI.

TINCTURE OF BLACK HENBANE.

Take of

Leaves of black henbane dried, one ounce,

Diluted alcohol, eight ounces.

Digest for seven days and filtrate through paper.

TINCTURA KINO.

TINCTURE OF KINO.

Take of

Kino in powder, two ounces,

Diluted alcohol, a pound and a half.

Digest seven days and filtrate through paper.

TINCTURA LAURI CINNAMOMI.

TINCTURE OF CINNAMON.

Take of

Bark of cinnamon bruised, three ounces,

Diluted alcohol, two pounds and a half.

Digest for seven days and filtrate through paper.

TINCTURA CINNAMOMI COMPOSITA.

COMPOUND TINCTURE OF CINNAMON.

Take of

Bark of cinnamon,

Lesser cardamom seeds, each bruised, one ounce,

Long pepper in powder, two drachms,

Diluted alcohol, two pounds and a half.

Digest for seven days and filtrate through paper.

TINCTURA LAVANDULÆ COMPOSITA.

COMPOUND TINCTURE OF LAVENDER.

Take of

Spirit of lavender, three pounds,

——— rosemary, one pound,

Bark of cinnamon bruised, one ounce,

Flower-buds of the clove tree bruised, two drachms,

Nutmeg bruised, half an ounce,

Wood of red saunders rasped, three drachms.

Macerate for seven days and filtrate.

TINCTURA MELOES VESICATORII.

TINCTURE OF CANTHARIDES.

Take of

Cantharides bruised, one drachm,

Diluted alcohol, one pound.

Digest for seven days and filtrate through paper.

TINCTURA MELOES VESICATORII FORTIOR.

STRONG TINCTURE OF CANTHARIDES.

Take of

Cantharides bruised, ten drachms,

Diluted alcohol, one pound.

Digest for fourteen days and filtrate through paper.

TINCTURA MIMOSÆ CATECHU.

TINCTURE OF CATECHU.

Take of

Extract of catechu in powder, three ounces,

Bark of cinnamon bruised, two ounces,

Diluted alcohol, two pounds and a half.

Digest for eight days and filtrate through paper.

TINCTURA MOSCHI.

TINCTURE OF MUSK.

Take of

Musk, two drachms,

Alcohol, one pound.

Macerate for seven days and filtrate.

TINCTURA MURIATIS AMMONIÆ ET FERRI.

TINCTURE OF MURIATE OF AMMONIA AND IRON.

Take of

Muriate of ammonia and iron, four ounces,
Diluted alcohol, sixteen ounces.

Digest and filtrate.

TINCTURA MURIATIS FERRI.

TINCTURE OF MURIATE OF IRON.

Take of

Carbonate of iron, half a pound,
Muriatic acid, three pounds,
Alcohol, three pounds and four ounces.

Pour the muriatic acid on the carbonate of iron in a glass vessel ; and shake the mixture now and then, during three days. Set it by, that the feces may subside ; then pour off the liquor ; evaporate this to sixteen ounces, and when cold, add to it the alcohol.

TINCTURA MYRRHÆ.

TINCTURE OF MYRRH.

Take of

Myrrh in powder, three ounces,
Alcohol, twenty ounces,
Water, ten ounces.

Digest for seven days and filtrate through paper.

TINCTURA OPII.

TINCTURE OF OPIUM.

Take of

Opium, two ounces,

Diluted alcohol, two pounds.

Digest seven days and filtrate through paper.

TINCTURA OPII CAMPHORATA.

CAMPHORATED TINCTURE OF OPIUM.

Take of

Opium,

Benzoic acid, of each one drachm,

Camphor, two scruples,

Volatile oil of aniseed, one drachm,

Diluted alcohol, two pounds.

Digest for ten days and filtrate through paper.

TINCTURA RHEI PALMATI.

TINCTURE OF RHUBARB.

Take of

Root of rhubarb in coarse powder, three ounces,

Lesser cardamom seeds bruised, half an ounce,

Diluted alcohol, two pounds and a half.

Digest for seven days and filtrate through paper.

TINCTURA RHEI ET ALOES.

TINCTURE OF RHUBARB AND ALOES.

Take of

Root of rhubarb in coarse powder, ten drachms,

Socotorine aloes in powder, six drachms,

Lesser cardamom seeds bruised, half an ounce,

Diluted alcohol, two pounds and a half.

Digest for seven days and filtrate through paper.

TINCTURA RHEI ET GENTIANÆ.

TINCTURE OF RHUBARB AND GENTIAN.

Take of

Root of rhubarb in coarse powder, two ounces,

Root of gentian sliced and bruised, half an ounce,

Diluted alcohol, two pounds and a half.

Digest for seven days and filtrate through paper.

TINCTURA RHÆI DULCIS.

SWEET TINCTURE OF RHUBARB.

Take of

Root of rhubarb in coarse powder, two ounces,

Root of liquorice bruised,

Anise seeds bruised, each one ounce,

Double refined sugar in powder, two ounces,

Diluted alcohol, two pounds and a half.

Digest for seven days and filtrate through paper.

TINCTURA SAPONIS.

TINCTURE OF SOAP.

Take of

Soap in shavings, four ounces,

Camphor, two ounces,

Volatile oil of rosemary, half an ounce,

Alcohol, two pounds.

Digest the soap in the alcohol for three days; then add to the filtrated liquor, the camphor and oil, agitating them diligently.

TINCTURA SAPONIS ET OPII.

TINCTURE OF SOAP AND OPIUM.

This is prepared in the same way and from the same substances as the tincture of soap, but with the addition from the beginning of one ounce of opium.

TINCTURA SCILLÆ MARITIMÆ.

TINCTURE OF SQUILL.

Take of

Root of squills fresh dried and bruised, four ounces,

Diluted alcohol, two pounds.

Digest for eight days and pour off the liquor.

TINCTURA TOLUIFERÆ BALSAMI.

TINCTURE OF BALSAM OF TOLU.

Take of

Balsam of tolu, an ounce and a half,

Alcohol, one pound.

Digest until the balsam be dissolved, and then filtrate the tincture through paper.

TINCTURA VERATRI ALBI.

TINCTURE OF WHITE HELLEBORE.

Take of

Root of white hellebore bruised, eight ounces,

Diluted alcohol, two pounds and a half.

Digest for seven days and filtrate through paper.

TINCTURES MADE WITH ETHEREAL SPIRITS.

TINCTURA ALOES ÆTHEREA.

ETHERIAL TINCTURE OF ALOES.

Take of

Gum resin of socotorine aloes,

Myrrh, of each in powder an ounce and a half,

English saffron sliced, one ounce,

Sulphuric ether with alcohol, one pound.

Digest the myrrh with the liquor for four days in a close vessel, then add the saffron and aloes.

Digest again for four days, and when the feces have subsided, pour off the tincture.

ÆTHER SULPHURICUS CUM ALCOHOLE AROMATICUS.

AROMATIC SULPHURIC ETHER WITH ALCOHOL.

This is made of the same aromatics, and in the same manner as the compound tincture of cinnamon; except that in place of the alcohol, sulphuric ether with alcohol is employed.

ACIDUM SULPHURICUM AROMATICUM.

AROMATIC SULPHURIC ACID.

Take of

Alcohol, two pounds.

Sulphuric acid, six ounces.

Drop the acid gradually into the alcohol.

Digest the mixture with a very gentle heat in a close vessel for three days, and then add of

Bark of cinnamon bruised, an ounce and a half,

Root of ginger bruised, one ounce.

Digest again in a close vessel for six days, and then filtrate the tincture through paper placed in a glass funnel.

AMMONIATED OR VOLATILE TINCTURES.

ALCOHOL AMMONIATUM AROMATICUM.

AROMATIC AMMONIATED ALCOHOL.

Take of

Ammoniated alcohol, eight ounces,

Volatile oil of rosemary, one drachm and a half,

Volatile oil of lemon peel, one drachm.

Mix them, that the oils may be dissolved.

TINCTURA CASTOREI COMPOSITA.

COMPOUND TINCTURE OF CASTOR.

Take of

Russian castor in powder, one ounce,

Gum resin of assa fœtida, half an ounce,

Ammoniated alcohol, one pound.

Digest for seven days in a close stopped phial, and filtrate through paper.

TINCTURA GUAJACI AMMONIATA.

AMMONIATED TINCTURE OF GUIACUM.

Take of

Gum resin of officinal guaiacum, four ounces,

Ammoniated alcohol, one pound and a half.

Digest for seven days and filtrate through paper.

TINCTURA OPII AMMONIATA.

AMMONIATED TINCTURE OF OPIUM.

Take of

Benzoic acid,
English saffron sliced, each three drachms,
Opium, two drachms,
Volatile oil of anise seeds, half a drachm,
Ammoniated alcohol, sixteen ounces.

Digest for seven days in a close vessel, and filtrate through paper.

CHAP. XV.

MEDICATED WINES.

VINUM ALOES SOCOTORINÆ.

WINE OF SOCOTORINE ALOES.

Take of

Gum resin of socotorine aloes in powder, one ounce,

Lesser cardamom seeds bruised,

Root of ginger bruised, each one drachm,

Spanish white wine, two pounds.

Digest for seven days, stirring now and then, and afterwards filtrate.

VINUM FERRI.

WINE OF IRON.

Take of

Purified filings of iron, one ounce,

Spanish white wine, sixteen ounces.

Digest for a month, often shaking the vessel, and then filtrate.

VINUM GENTIANÆ COMPOSITUM.

COMPOUND WINE OF GENTIAN.

Take of

Root of gentian sliced and bruised, half an ounce,

Red bark of cinchona in powder, one ounce,

External rind of seville oranges dried and bruised, two drachms,

Bark of canella alba in powder, one drachm,

Diluted alcohol, four ounces,

Spanish white wine, two pounds and a half.

First, pour on the diluted alcohol, and after twenty-four hours, add the wine ; then macerate for seven days and filtrate.

VINUM IPECACUANHÆ.

WINE OF IPECACUAN.

Take of

Root of ipecacuan in powder, one ounce,

Spanish white wine, fifteen ounces.

Macerate for seven days and filtrate through paper.

VINUM NICOTIANÆ TABACI.

WINE OF TOBACCO.

Take of

Dried leaves of tobacco sliced, one ounce,
Spanish white wine, one pound.

Macerate for seven days and filtrate through paper.

VINUM OPII COMPOSITUM.

COMPOUND WINE OF OPIUM.

Take of

Opium, two ounces,
Flower buds of the clove tree bruised,
Cinnamon bruised, each one drachm,
Spanish white wine, sixteen ounces.

Macerate for seven days and filtrate through paper.

VINUM RHEI PALMATI.

WINE OF RHUBARB.

Take of

Root of rhubarb sliced, two ounces,
Bark of canella alba bruised, one drachm,
Diluted alcohol, two ounces,
Spanish white wine, fifteen ounces.

Macerate for seven days and filtrate through paper.

VINUM TARTRITIS ANTIMONII.

WINE OF TARTRITE OF ANTIMONY.

Take of

Tartrite of antimony, twenty-four grains,
Spanish white wine, one pound.

Mix them so that the tartrite of antimony may be dissolved.

CHAP. XVI.

EXTRACTS.

EXTRACTS MADE WITH WATER.

EXTRACTUM GENTIANÆ LUTEÆ.

EXTRACT OF GENTIAN.

Take of

Root of gentian, any quantity.

Having cut and bruised it, pour upon it eight times its quantity of water. Boil to the consumption of one half of the liquor, and strain it by strong expression. Evaporate the decoction immediately to the consistence of thick honey, in a bath of water saturated with muriate of soda.

In the same manner are prepared

From the roots of

Liquorice, *Extractum glycyrrhizæ glabræ.*
Extract of liquorice.

Black hellebore, *Extractum hellebori nigri.*
Extract of black hellebore.

From the inner bark of

Butternut, *Extractum juglandis cinereæ.*
Extract of butternut.

From the leaves of

Rue, *Extractum rutæ graveolentis.*
Extract of rue.

Senna, *Extractum cassiæ sennæ.*
Extract of senna.

From the flowers of

Chamomile, *Extractum anthemidis nobilis.*
Extract of chamomile.

From the capsules of

White poppy, *Extractum papáveris somniferi.*
Extract of white poppy.

From the wood of

Logwood, *Extractum hæmatoxyli campechensis.*
Extract of logwood.

EXTRACTS MADE WITH ALCOHOL AND WATER.

EXTRACTUM CINCHONÆ OFFICINALIS.

EXTRACT OF CINCHONA.

Take of

Bark of cinchona in powder, one pound,
Alcohol, four pounds.

Digest for four days, and pour off the tincture. Boil the residuum in five pounds of distilled water for fifteen minutes, and filtrate the decoction boiling hot through linen. Repeat this decoction and filtration with an equal quantity of distilled water, and reduce the liquor by evaporation to the consistence of thin honey. Draw off the alcohol from the tincture by distillation, until it also become thick; then mix the liquors thus inspissated, and evaporate in a bath of boiling water, saturated with muriate of soda, to a proper consistency.

EXTRACTUM CONVOLVULI JALAPÆ.

EXTRACT OF JALAP.

This is prepared in the same way.

CHAP. XVII.

OF POWDERS.

PULVIS CINNAMOMI COMPOSITUS.

COMPOUND POWDER OF CINNAMON.

Take of

Bark of cinnamon,
Lesser cardamom seeds,
Root of ginger, each equal parts.

Reduce them to a very fine powder, which is to be kept in a glass vessel well closed.

PULVIS ASARI COMPOSITUS.

COMPOUND POWDER OF ASARABACCA.

Take of

Leaves of asarabacca, three parts,
——— sweet marjorum,
Flowers of lavender, each one part.

Rub them together to a powder.

PULVIS CARBONATIS CALCIS COMPOSITUS.

COMPOUND POWDER OF CARBONATE OF LIME.

Take of

Prepared carbonate of lime, four ounces,

Nutmeg, half a drachm,

Bark of cinnamon, one drachm and a half.

Reduce them together to powder.

PULVIS IPECACUANHÆ ET OPII.

POWDER OF IPECACUAN AND OPIUM.

Take of

Root of ipecacuan in powder,

Opium, of each one part,

Sulphate of potass, eight parts.

Triturate them together into a fine powder.

PULVIS JALAPÆ COMPOSITUS.

COMPOUND POWDER OF JALAP.

Take of

Root of jalap, one part,

Super-tartrite of potass, two parts.

Grind them together to a very fine powder.

PULVIS OPIATUS.

OPIATE POWDER.

Take of

Opium, one part,

Prepared carbonate of lime, nine parts.

Rub them together to a very fine powder.

PULVIS SCAMMONII COMPOSITUS.

COMPOUND POWDER OF SCAMMONY.

Take of

Gum resin of scammony,

Super-tartrite of potass, equal parts.

Rub them together to a very fine powder.

PULVIS SUPER-SULPHATIS ALUMINÆ ET POTASSÆ COMPOSITUS.

COMPOUND POWDER OF SUPER-SULPHATE OF ALUMINA AND POTASS.

Take of

Super-sulphate of alumina and potass, four parts,

Kino, one part.

Rub them together to a fine powder.

CHAP. XVIII.

CONSERVES AND ELECTUARIES.

Conservæ	*Conserves*
Corticis exterioris recentis fructus citri aurantii radulâ abrasi.	Of the outer rind of oranges rasped off by a grater.
Petalorum rosæ gallicæ nondum explicitorum.	Of red rose buds.

Beat each of these to a pulp, gradually adding during the beating, three times the weight of double refined sugar.

ELECTUARIUM AROMATICUM.

AROMATIC ELECTUARY.

Take of

Compound powder of cinnamon, one part,

Syrup of orange peel, two parts.

Mix and beat them well together, so as to form an electuary.

ELECTUARIUM CASSIÆ FISTULÆ.

ELECTUARY OF CASSIA.

Take of

Pulp of the fruit of the cassia tree, six ounces,
Pulp of tamarinds,
Manna, each an ounce and a half,
Syrup of damask roses, six ounces.

Having beat the manna in a mortar, dissolve it in a gentle heat in the syrup; then add the pulps and evaporate them with a regular continued heat, to the consistence of an electuary.

ELECTUARIUM CASSIÆ SENNÆ.

ELECTUARY OF SENNA.

Take of

Leaves of senna, eight ounces,
Coriander seeds, four ounces,
Root of liquorice, three ounces,
Figs,
Pulp of prunes, each one pound,
—— of tamarinds, half a pound,
Double refined sugar, two pounds and a half.

Powder the senna with the coriander seeds, and

sift out ten ounces of the mixed powder. Boil the remainder with the figs and liquorice in four pounds of water, to one half; then press out and strain the liquor. Evaporate this strained liquor to the weight of about a pound and a half; then add the sugar, and make a syrup; add this syrup by degrees to the pulps, and, lastly, mix in the powder.

ELECTUARIUM CATECHU.

ELECTUARY OF CATECHU.

Take of

Extract of catechu, four ounces,
Kino, three ounces,
Bark of cinnamon,
Nutmeg, each one ounce,
Opium diffused in a sufficient quantity of spanish white wine, one drachm and a half,
Syrup of red roses boiled to the consistence of honey, two pounds and a quarter.

Reduce the solids to powder; and having mixed them with the opium and syrup, make them into an electuary.

ELECTUARIUM OPIATUM.

OPIATE ELECTUARY.

Take of

Compound powder of cinnamon, six ounces,
Virginian snakeroot in fine powder, three ounces,
Opium diffused in a sufficient quantity of spanish white wine, half an ounce,
Syrup of ginger, one pound.

Mix them and form an electuary.

CHAP. XIX.

TROCHES.

TROCHISCI CARBONATIS CALCIS.

TROCHES OF CARBONATE OF LIME.

Take of

Prepared carbonate of lime, four ounces,
Gum arabic, one ounce,
Nutmeg, one drachm,
Double refined sugar, six ounces.

Powder them together and form them with water into a mass, which is to be made into troches.

TROCHISCI GLYCYRRHIZÆ CUM OPIO.

TROCHES OF LIQUORICE WITH OPIUM.

Take of

Opium, two drachms,
Tincture of balsam of tolu, half an ounce,
Simple syrup, eight ounces,
Extract of liquorice, softened in warm water,
Gum arabic in powder, each five ounces.

Grind the opium well with the tincture, then

add by degrees the syrup and extract ; afterwards gradually sprinkle upon the mixture the powdered gum arabic. Lastly, dry them so as to form a mass to be made into troches, each weighing ten grains.

TROCHISCI MAGNESIÆ.

TROCHES OF MAGNESIA.

Take of

Magnesia, four ounces,
Double refined sugar, two ounces,
Root of ginger powdered, one scruple.

Triturate them together, and with the addition of the mucilage of gum arabic, make troches.

CHAP XX.

PILLS.

PILULÆ ALOETICÆ.

ALOETIC PILLS.

Take of

Gum resin of socotorine aloes in powder,

Soap, equal parts.

Beat them with simple syrup into a mass fit for making pills.

PILULÆ ALOES ET ASSÆ FŒTIDÆ.

PILLS OF ALOES AND ASSA FOETIDA.

Take of

Gum resin of socotorine aloes in powder,

Gum resin of assa fœtidâ,

Soap, equal parts.

Form them into a mass with mucilage of gum arabic.

PILULÆ ALOES ET COLOCYNTHIDIS.

PILLS OF ALOES AND COLOQUINTIDA.

Take of

Gum resin of socotorine aloes,
———— scammony, each eight parts,
Fruit of coloquintida, four parts,
Volatile oil of cloves,
Sulphate of potass with sulphur, each one part.

Reduce the aloes and scammony into a powder with the salt; then let the fruit of coloquintida beat into a very fine powder, and the oil be added; lastly, make it into a proper mass with the mucilage of gum arabic.

PILULÆ ALOES ET MYRRHÆ.

PILLS OF ALOES AND MYRRH.

Take of

Gum resin of socotorine aloes, two ounces.
Myrrh, one ounce,
Saffron, half an ounce.

Beat them into a mass with a proper quantity of simple syrup.

PILULÆ ASSÆ FŒTIDÆ COMPOSITÆ.

COMPOUND PILLS OF ASSA FOETIDA.

Take of

Gum resin of assa fœtida,

Galbanum,

Myrrh, each eight parts,

Purified oil of amber, one part.

Beat them into a mass with simple syrup.

PILULÆ AMMONIARETI CUPRI.

PILLS OF AMMONIARET OF COPPER.

Take of

Ammoniaret of copper in fine powder, sixteen grains,

Bread crumb, four scruples,

Solution of carbonate of ammonia, as much as may be sufficient.

Beat them into a mass, to be divided into thirty-two equal pills.

PILULÆ HYDRARGYRI.

PILLS OF QUICKSILVER.

Take of

Purified quicksilver,
Conserve of red roses, each one ounce,
Starch, two ounces.

Triturate the quicksilver with the conserve in a glass mortar, till the globules completely disappear, adding occasionally a little mucilage of gum arabic; then add the starch, and beat the whole with water into a mass, which is immediately to be divided into four hundred and eighty equal pills.

PILULÆ OPIATÆ.

PILLS OF OPIUM.

Take of

Opium, one part,
Extract of liquorice, seven parts,
Jamaica pepper, two parts.

Soften the opium and extract separately, with diluted alcohol, and having beat them into a pulp, mix them; then add the pepper reduced to powder; and lastly, having beat them well together, form the whole into a mass.

PILULÆ RHEI COMPOSITÆ.

COMPOUND PILLS OF RHUBARB.

Take of

Root of rhubarb in powder, one ounce,
Gum resin of socotorine aloes, six drachms,
Myrrh, half an ounce,
Volatile oil of peppermint, half a drachm.

Make them into a mass with a sufficient quantity of syrup of orange peel.

PILULÆ SCILLITICÆ.

SQUILL PILLS.

Take of

Dried root of squills in fine powder, one scruple,
Ammoniacum,
Lesser cardamom seeds in powder,
Extract of liquorice, each one drachm.

Mix and form them into a mass with simple syrup.

CHAP. XXI.

LINIMENTS, OINTMENTS, CERATES, AND PLASTERS.

In making these compositions, the fatty and resinous substances are to be melted with a gentle heat, and then constantly stirred, adding at the same time the dry ingredients, if there be any, until the mixture on cooling become stiff.

LINIMENTUM SIMPLEX.

SIMPLE LINIMENT.

Take of

Olive oil, four parts,
White wax, one part.

UNGUENTUM SIMPLEX.

SIMPLE OINTMENT.

Take of

Olive oil, five parts,
White wax, two parts.

UNGUENTUM ROSARUM.

OINTMENT OF ROSES.

Take of

Hog's lard,

Fresh damask roses with their calices, each equal parts.

Let the roses be slightly bruised in a marble mortar with a pestle of wood, and put them in a vessel with the lard; place this over a gentle fire, so as to evaporate a great part of the moisture; then press it through linen and suffer it to cool. Separate the feces which are on the top, and melt it in order to depurate.

UNGUENTUM AQUÆ ROSÆ.

OINTMENT OF ROSE WATER.

Take of

Oil of sweet almonds, two ounces,

Spermaceti, half an ounce,

White wax, one drachm.

Melt the whole in a water bath, stirring it frequently; when melted, add of

Damask rose water, two ounces,

and stir the mixture continually till it is cold.

UNGUENTUM STRAMONII.

OINTMENT OF THORN APPLE.

Take of

Leaves of thorn apple recently gathered and sliced, five pounds,

Hog's lard, fourteen pounds.

Let them simmer together over a gentle fire till the leaves become crisp and dry. Then press out the lard, return it into the vessel when cleansed, and add to every pound of the compound, of

Yellow wax, two ounces.

Set the whole on the fire ; when the wax has melted remove the vessel, and let it rest while the contents gradually cool, that the impurities may subside. These must then be separated from the ointment.

UNGUENTUM JUNIPERI COMMUNIS.

OINTMENT OF JUNIPER.

Take of

Leaves of juniper, recently gathered and sliced,

Yellow wax, each one pound,

Hog's lard, two pounds.

Mix the articles, and melt ; boil for a short

time, taking care to avoid empyreuma. Strain while the mixture is hot through a coarse sieve.

UNGUENTUM RESINOSUM.

RESINOUS OINTMENT.

Take of

Hog's lard, eight parts,
Resin of pine, five parts,
Yellow wax, two parts.

UNGUENTUM PICIS.

TAR OINTMENT.

Take of

Tar, five parts,
Yellow wax, two parts.

UNGUENTUM INFUSI MELOES VESICATORII.

OINTMENT OF INFUSION OF CANTHARIDES.

Take of

Cantharides,
Resin of pine,
Yellow wax, each one part,
Hog's lard,
Venice turpentine, each two parts,
Boiling water, four parts.

Infuse the cantharides in the water for a night; then strongly press out and strain the liquor and boil it with the lard till the water be consumed; then add the resin and wax, and when these are melted, take the ointment off the fire and add the turpentine.

UNGUENTUM PULVERIS MELOES VESICATORII.

OINTMENT OF THE POWDER OF CANTHARIDES.

Take of

Resinous ointment, seven parts,
Powdered cantharides, one part.

UNGUENTUM SULPHURIS.

SULPHUR OINTMENT.

Take of

Hog's lard, two parts,
Sublimed sulphur, one part.

To each pound of this ointment, add of

Volatile oil of lemons, or
———— of lavender, half a drachm.

UNGUENTUM ACIDI NITROSI.

OINTMENT OF NITROUS ACID.

Take of

Hog's lard, one pound,
Nitrous acid, six drachms.

Mix the acid gradually with the melted lard, and diligently beat the mixture as it cools.

UNGUENTUM OXIDI PLUMBI ALBI.

OINTMENT OF WHITE OXIDE OF LEAD.

Take of

Simple ointment, five parts,
White oxide of lead, one part.

UNGUENTUM ACETITIS PLUMBI.

OINTMENT OF ACETITE OF LEAD.

Take of

Simple ointment, twenty parts,
Acetite of lead, one part.

UNGUENTUM HYDRARGYRI.

OINTMENT OF QUICKSILVER.

Take of

Quicksilver,
Hog s lard, each three parts,
Mutton suet, one part.

Rub the quicksilver carefully in a mortar with a little of the hog s lard, until the globules entirely disappear; then add the remainder of the lard and the suet, rubbing them well together.

UNGUENTUM HYDRARGYRI MITIUS.

MILD OINTMENT OF QUICKSILVER.

This is to be prepared as the preceding ointment, excepting that only one part of quicksilver is to be employed.

UNGUENTUM OXIDI HYDRARGYRI CINEREI.

OINTMENT OF GREY OXIDE OF QUICKSILVER.

Take of

Grey oxide of quicksilver, one part,
Hog's lard, three parts.

UNGUENTUM SUB-MURIATIS HYDRARGYRI ET AMMONIÆ.

OINTMENT OF SUB-MURIATE OF QUICKSILVER AND AMMONIA.

Take of

Sub-muriate of quicksilver and ammonia, one drachm,

Ointment of roses, one ounce and a half.

Let them be mixed very intimately.

UNGUENTUM OXIDI HYDRARGYRI RUBRI.

OINTMENT OF RED OXIDE OF QUICKSILVER.

Take of

Red oxide of quicksilver by nitrous acid, one part,

Hog's lard, eight parts.

UNGUENTUM NITRATIS HYDRARGYRI.

OINTMENT OF NITRATE OF QUICKSILVER.

Take of

Purified quicksilver, one part,

Nitrous acid, two parts,

Hog's lard, three parts,

Olive oil, nine parts.

Dissolve the quicksilver in the nitrous acid, by

digestion in a sand heat, and, while the solution is hot, mix with it the lard and oil, previously melted together, and just beginning to grow stiff. Stir them briskly together in a glass or wedgwood mortar, so as to form the whole into an ointment.

UNGUENTUM NITRATIS HYDRARGYRI MITIUS.

MILDER OINTMENT OF NITRATE OF QUICKSILVER.

This is prepared in the same way, with three times the quantity of hog's lard and olive oil.

UNGUENTUM SUB-ACETITIS CUPRI.

OINTMENT OF SUB-ACETITE OF COPPER.

Take of

Resinous ointment, fifteen parts,
Sub-acetite of copper, one part.

UNGUENTUM OXIDI ZINCI IMPURI.

OINTMENT OF IMPURE OXIDE OF ZINC.

Take of

Simple liniment, five parts,
Prepared impure oxide of zinc, one part.

UNGUENTUM OXIDI ZINCI.

OINTMENT OF OXIDE OF ZINC.

Take of

Simple liniment, six parts,
Oxide of zinc, one part.

CERATUM SIMPLEX.

SIMPLE CERATE.

Take of

Olive oil, six parts,
White wax, three parts,
Spermaceti, one part.

CERATUM CARBONATIS ZINCI IMPURI.

CERATE OE IMPURE CARBONATE OF ZINC.

Take of

Simple cerate, five parts,
Prepared impure carbonate of zinc, one part.

EMPLASTRUM SIMPLEX.

SIMPLE PLASTER.

Take of

Yellow wax, three parts,
Mutton suet,
Resin of pine, each two parts.

EMPLASTRUM RESINOSUM COMPOSITUM.

COMPOUND RESINOUS PLASTER.

Take of

Burgundy pitch, two pounds,
Galbanum, one pound,
Resin of pine,
Yellow wax, each four ounces,
Fixed oil of mace, one ounce.

To the pitch, resin and wax melted together, add first the galbanum and then the oil of mace.

EMPLASTRUM MELOES VESICATORII.

PLASTER OF SPANISH FLIES.

Take of

Mutton suet,
Yellow wax,
Resin of pine,
Cantharides, each equal weights.

Beat the cantharides into a fine powder, and add them to the other ingredients previously melted and removed from the fire.

EMPLASTRUM MELOES VESICATORII COMPOSITUM.

COMPOUND PLASTER OF SPANISH FLIES.

Take of

Venice turpentine, eighteen parts,
Burgundy pitch,
Cantharides, each twelve parts,
Yellow wax, four parts,
Sub-acetite of copper, two parts,
Mustard seed,
Black pepper, each one part.

Having first melted the pitch and wax, add the turpentine, and to these in fusion, and still hot, add the other ingredients, reduced to a fine powder and mixed, and stir the whole carefully together, so as to form a plaster.

EMPLASTRUM OXIDI PLUMBI SEMIVITREI.

PLASTER OF THE SEMI-VITRIFIED OXIDE OF LEAD.

Take of

Semi-vitrified oxide of lead, one part,
Olive oil, two parts.

Boil them, adding water, and constantly stirring the mixture till the oil and oxide be formed into a plaster.

EMPLASTRUM RESINOSUM.

RESINOUS PLASTER.

Take of

Plaster of semi-vitrified oxide of lead,
Resin of pine, equal weights.

EMPLASTRUM ASSÆ FŒTIDÆ.

PLASTER OF ASSA FOETIDA.

Take of

Plaster of semi-vitrified oxide of lead,
Gum resin of assa fœtida, each two parts,
Galbanum,
Yellow wax, each one part.

EMPLASTRUM GUMMOSUM.

GUM PLASTER.

Take of

Plaster of semi-vitrified oxide of lead, eight parts,
Ammoniacum,
Galbanum,
Yellow wax, each one part.

EMPLASTRUM SAPONACEUM.

SAPONACEOUS PLASTER.

Take of

Plaster of semi-vitrified oxide of lead, four parts,

Gum plaster, two parts,

Soap sliced, one part.

To the plasters melted together, add the soap; then boil for a little while so as to form a plaster.

EMPLASTRUM HYDRARGYRI.

PLASTER OF QUICKSILVER.

Take of

Olive oil,

Resin of pine, each one part,

Quicksilver, three parts,

Plaster of semi-vitrified oxide of lead, six parts.

Melt the oil and resin together, and when this mixture is cold let the quicksilver be rubbed with it till the globules disappear; then add by degrees the plaster of semi-vitrified oxide of lead melted, and let the whole be accurately mixed.

EMPLASTRUM OXIDI FERRI RÚBRI.

PLASTER OF RED OXIDE OF IRON.

Take of

Plaster of semi-vitrified oxide of lead, twenty-four parts,
Resin of pine, six parts,
Yellow wax,
Olive oil, each three parts,
Red oxide of iron, eight parts.

Grind the red oxide of iron with the oil, and then add it to the other ingredients, previously melted.

TABLES.

S

TABLE

SHEWING THE PROPORTION OF ANTIMONY, OPIUM, AND QUICKSILVER, CONTAINED IN SOME COMPOUND MEDICINES.

TARTRITE OF ANTIMONY.

WINE OF TARTRITE OF ANTIMONY contains two grains of tartrite of antimony in the ounce.

OPIUM.

OPIATE ELECTUARY contains in each drachm about a grain and a half of opium.

ELECTUARY OF CATECHU contains in each ounce about two grains and a half of opium; for one grain of opium is contained in one hundred and ninety-three grains.

POWDER OF IPECACUAN AND OPIUM contains six grains of opium in each drachm, or one grain in ten.

OPIATE POWDER contains one grain of opium in ten.

OPIATE PILLS contain six grains of opium in each drachm, or five grains contain half a grain of opium.

TINCTURE OF OPIUM is made with two scruples of opium in each ounce of the liquid, or with five grains in each drachm. But a drachm of the tincture appears, by evaporation, to contain about three grains and a half of opium.

AMMONIATED TINCTURE OF OPIUM is made with about eight grains of opium in each ounce of the liquid, or with about one grain in the drachm.

TINCTURE OF SOAP WITH OPIUM is made with about fifteen grains of opium in each ounce of the liquid.

TROCHES OF LIQUORICE WITH OPIUM contain about one grain of opium in each drachm.

CAMPHORATED TINCTURE OF OPIUM contains nearly one grain of opium in three drachms.

QUICKSILVER.

QUICKSILVER PILLS contain fifteen grains of quicksilver in each drachm. Each pill contains one grain of quicksilver.

QUICKSILVER OINTMENT contains about twenty-five grains of quicksilver in each drachm.

MILD QUICKSILVER OINTMENT contains twelve grains of quicksilver in each drachm.

QUICKSILVER PLASTER contains about sixteen grains of quicksilver in each drachm.

OINTMENT OF NITRATE OF QUICKSILVER contains in each drachm four grains of quicksilver and eight of nitrous acid.

MILDER OINTMENT OF NITRATE OF QUICKSILVER contains in each scruple half a grain of quicksilver, and one grain of nitrous acid.

OINTMENT OF THE SUB-MURIATE OF QUICKSILVER AND AMMONIA contains in each drachm about four and a half grains of the oxyd.

POSOLOGICAL AND PROSODIAL TABLE.

Acetītis ammonĭæ aqua, ʒ ij ad vi.

Acĭdum acetōsum, ʒ i ad ℥ ss.

——————— destillātum, idem.

——————— forte, ℈ ad ʒ i.

——— benzoĭcum, gr. x a ʒ ss.

Acĭdi carbōnĭci aqua, ℔ ij.

Acidum muriāticum, gt. x ad xl.

——— nitrōsum, gt. v ad xx.

——————— dilūtum, gt. x ad xl.

——— succĭnĭcum, gr. v ad ℈ i.

——— sulphŭrĭcum dilūtum, gt. xv ad l.

——————— aromăticum, gt. xv ad l.

Aconīti neomōntani herba, gr. i ad v.

——————— succus spissātus, gr. ½ ad iii.

Acōri călămi rādix, ℈ i ad ʒ i.

Aescŭli hippocastăni cortex, ʒ ss ad i.

Aether sulphurĭcus, gt. xx ad ʒ i.

——————— cum alcohōle, ʒ ss ad ij.

Aether sulphurĭcus cum alcohōle, aromatĭcus, ʒ ss ad ij.

Alcŏhol, ʒ ss ad i.

——— ammoniātum, ʒ ss ad i.

——————— aromatĭcum, ʒ ss ad i.

——————— fœtidum, ʒ ss ad i.

Allĭi satīvi rādix, ʒ i ad ij.

Alŏës perfoliātæ socotorīnæ gummi-resīna, gr. v ad xv.

————— pilŭlæ, gr. xv ad ʒ ss.

————— et assæ fœtidæ pilŭlæ, gr. x ad ℈ i.

————— et cŏlŏcynthĭdis pilŭlæ, gr. v ad x.

————— et myrrhæ pilŭlæ, gr. x ad ℈ i.

————— tinctūra, ʒ ss ad ij.

————— et myrrhæ tinctūra, ʒ ss ad ij.

————— tinctura ætherĕa, ʒ ss ad ij.

————— vinum, ℥ ss ad iss.

————— syrūpus, ʒ i ad iij.

Alūmĭnæ et potassæ super-sulphas, ℈ ss ad i.

——— et potassæ super-sulphātis pulvis composĭtus, gr. x ad ʒ ss.

Ammōniæ aqua, gt. x ad xxx.

———— acetītis aqua, ℥ ss.

———— hydro-sulphurētum, gt. v ad xij.

———— carbōnas, gr. v ad xv.

———— carbōnātis aqua, gt. xx ad ʒ i.

Ammōniăcum gummi-resīna, gr. x ad ʒ ss.
Ammōniăci emulsio, ʒ iij ad ℥ i.
Amōmi zingĭbĕris rādix, gr. v ad ℈ i.
——————— syrūpus, ʒ i ad iij.
——— repentis semĭna, gr. v ad ℈ i.
——————— tinctūra, ʒ i ad iij.
——— zedoārĭæ rādix, ℈ i ad ʒ i.
Amygdăli commūnis oleum, ʒ iij ad ℥ i.
——————— emulsio, ℔ ij.
Amȳrĭdis gileadensis resīna liquida, ℈ i ad ʒ i.
Anēthi grăvĕŏlentis semĭna, ℈ i ad ʒ i.
—— fœnĭcŭli semĭnā, ℈ i ad ʒ i.
——————— oleum volatĭle, gt. ij ad v.
Angĕlĭcæ archangĕlĭcæ rādix, herba, semen, ʒ ss ad iss.
Angustūræ cortex, gr. x ad ℈ i.
Anthemĭdis nobĭlis flores, ℈ i ad ʒ i.
——————— extractum, gr. x ad ʒ i.
——————— pyrĕthri rādix, gr. iii. ad ℈ i.
Antimōnii oxĭdum cum phosphāte calcis, gr. iij ad xv.
——————— cum sulphŭre per nitrātem potassæ, gr. i ad iv.
——————— cum sulphŭre vitrificātum, gr. ¼ ad iss.

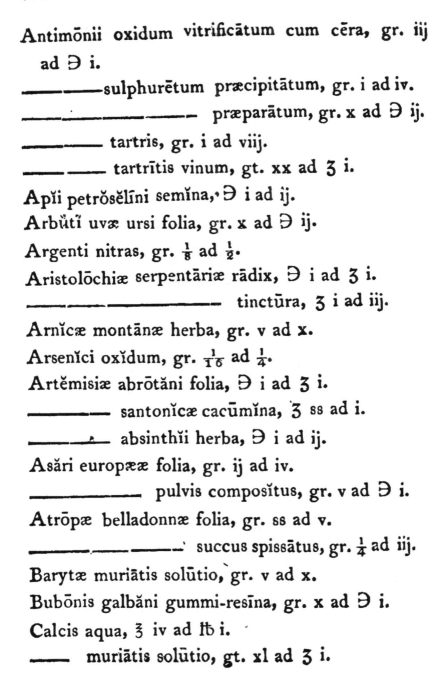

Antimōnii oxidum vitrificātum cum cēra, gr. iij ad ℈ i.

——— sulphurētum præcipitātum, gr. i ad iv.

——— præparātum, gr. x ad ℈ ij.

——— tartris, gr. i ad viij.

——— tartrītis vinum, gt. xx ad ʒ i.

Apĭi petrŏsĕlīni semĭna, ℈ i ad ij.

Arbŭtĭ uvæ ursi folia, gr. x ad ℈ ij.

Argenti nitras, gr. $\frac{1}{8}$ ad $\frac{1}{2}$.

Aristolōchiæ serpentāriæ rādix, ℈ i ad ʒ i.

——— tinctūra, ʒ i ad iij.

Arnĭcæ montānæ herba, gr. v ad x.

Arsenĭci oxĭdum, gr. $\frac{1}{16}$ ad $\frac{1}{4}$.

Artĕmisiæ abrōtăni folia, ℈ i ad ʒ i.

——— santonĭcæ cacūmĭna, ʒ ss ad i.

——— absinthĭi herba, ℈ i ad ij.

Asări europææ folia, gr. ij ad iv.

——— pulvis composĭtus, gr. v ad ℈ i.

Atrōpæ belladonnæ folia, gr. ss ad v.

——— succus spissātus, gr. $\frac{1}{4}$ ad iij.

Barytæ muriātis solūtio, gr. v ad x.

Bubōnis galbăni gummi-resīna, gr. x ad ℈ i.

Calcis aqua, ℥ iv ad ℔ i.

——— muriătis solūtio, gt. xl ad ʒ i.

Calcis carbōnas præparātus, ℈ i ad ʒ i.
—— carbōnātis mistūra, ℥ i ad ij.
———— pulvis composĭtus, ℈ i ad ij.
———— trochisci, ʒ i ad ij.
Canellæ albæ cortex, gr. xv ad ℈ ij.
Capsĭci annui fructus, gr. v ad x.
Cari carŭi semĭna, gr. x ad ʒ i.
——— spirĭtus, ʒ ij ad ℥ i.
Cassiæ fistŭlæ pulpa, ℥ ss ad i.
——— electuārium, ʒ ij ad ℥ i.
—— sennæ folia, ℈ i ad ʒ i.
——— tinctūra composita, ℥ ss ad i.
——— electuārium, ʒ i ad ℥ ss.
——— extractum, gr. x ad ʒ ss.
Castŏrĕum, gr. x ad ℈ i.
Castorĕi tinctūra, ʒ i ad ij.
———— composĭta, ʒ ss ad i.
Centaureæ benĕdictæ herba, gr. xv ad ʒ i.
Chironĭæ centaurĕi summitātes, ℈ i ad ʒ i.
Cinchōnæ officinālis cortex, ℈ i ad ʒ ij.
———— decoctum, ℥ i ad iv.
———— infūsum, ℥ i ad iv.
———— tinctūra, ʒ i ad ij.
———— tinctūra composĭta, ʒ i ad iij.
———— extractum, gr. x ad ℈ i.

Citri aurantĭi folia, flores, gr. x ad ʒ i.
———— fructûs cortex exterior, ʒ ss ad ℈ ij.
———— aqua destillāta, ℥ i ad iij.
———— syrūpus cortĭcis, ʒ i ad ij.
———— conserva cortĭcis, ʒ ij ad v.
—— medĭcæ succus spissātus, ʒ i ad ℥ ij.
———— syrūpus sūcci, ʒ i ad iij.
———— aqua destillāta, ℥ i ad iij.
Cochlearĭæ officinālis succus compositus, ℥ i ad iv.
———— armoracĭæ rādix, ℈ i ad ʒ i.
Colchici autumnālis rādix, gr. ss ad iij.
Cŏlombæ rādix, gr. x ad ℈ i.
———— tinctūra, ʒ i ad iij.
Conĭi maculati folia, gr. iij.
———— succus spissātus, gr. ½ ad gr. iij.
Convolvŭli scammōnĭæ gummi-resīna, gr. v ad xv.
———— pulvis compositus, gr. x ad xv.
———— electuārium, ʒ ss ad i.
———— jalapæ rādix, gr. x ad ʒ ss.
———— pulvis compositus, ʒ ss ad i.
———— tinctūra, ʒ i ad iij.
———— extractum, ℈ ss ad i.
Copaifĕræ officinālis resīna, gt. xv ad ʒ ss.
Corĭandri satīvi semĭna, ℈ i ad ʒ i.
Croci satīvi floris stigmăta, gr. v ad ʒ ss.

Crotōnis eleutheriæ cortex, ℈ i ad ʒ ss.
————— tinctūra, ʒ i ad ℥ ss.
Cucŭmis cŏlŏcynthĭdis fructūs medulla, gr. iij ad viij.
Cumīni cymīni semĭna, ℈ i ad ʒ i.
Cupri sub-acētis, gr. $\frac{1}{8}$ ad $\frac{1}{2}$.
—— ammoniarētum, gr. $\frac{1}{2}$ ad v.
—— ammoniarēti pilŭlæ, No. i.
—— sulphas, gr. ij ad x.
Curcŭmæ longæ rādix, ℈ i ad ʒ i.
Daphnes mezerēi rādīcis cortex, gr. i ad x.
————— decoctum, ℥ iv ad ℔ ss.
Datūræ stramonĭi folia, semina, gr. i ad v.
————— succus spissātus, gr. i ad x.
Dauci carotæ semĭna, ℈ i ad ʒ i.
Delphinĭi staphisāgrĭæ semĭna, gr. iij ad x.
Digitālis purpŭreæ folia, gr. ss ad iij.
————— infūsum, ʒ iij ad ℥ i.
————— tinctūra, gt. x ad xl.
Dōlĭchi prurĭentis pubes leguminis rigida, gr. v ad x.
Dorstēnĭæ contrajērvæ rādix, ℈ i ad ʒ ss.
Electuarium opiātum, ℈ i ad ij.
Eugeniæ caryophyllātæ floris germen, gr. v ad ℈ j.
————— oleum volatĭle, gt. ii. ad v.
Ferri limatūra purificāta, gr. iii ad x.
—— oxĭdum nigrum purificātum, *idem*.

Ferri carbōnas, gr. iii ad x.
—————— præcipitātus, *idem*.
—— super-carbōnātis aqua, ℔ i.
—— sulphas, gr. i ad v.
—— et ammōniæ murĭas, gr. iii ad xv.
—— murĭatis tinctūra, gt. x ad xx.
—————— et ammōniæ tinctūra, gt. xv ad ʒ i.
—— vinum, ʒ ij ad vi.
Ferŭlæ assæ fœtĭdæ gummi-resīna, gr. x ad ʒ ss.
—————————— tinctūra, ʒ ss ad i.
—————————— pilŭlæ composĭtæ, gr. x ad xx.
Fraxĭni orni succus concrētus, ℥ ss ad iss.
Gambōgia, gr. v ad x.
Gentiānæ lutĕæ rādix, gr. x ad ℈ ij.
—————— infūsum composĭtum, ℥ ss ad ij.
—————— tinctūra composĭta, ʒ i ad iii.
—————— vinum composĭtum, ℥ ss ad i.
—————— extractum, gr. x ad ℈ ij.
Geoffrææ inermis cortex, ℈ i ad ij.
Glycyrrhīzæ glabræ rādix, ʒ ss ad i.
—————————— extractum, ʒ i ad iij.
—————————— trochisci cum opio, ʒ ss ad j.
Gratiŏlæ officinālis herba, gr. x ad ℈ i.
Guāiăci officinālis resīna, gr. x ad ʒ ss.
—————————— tinctūra, ʒ ij ad ℥ ss.

Guāiăci officinālis tinctūra ammōniăta, ʒ i ad ij.
———————— decoctum compositum, ℥ iv ad vi.
Hæmatoxȳli campechiāni extractum, ℈ i ad ij.
Hellĕbŏri nigri rādix, gr. x ad ℈ i.
——————— extractum, gr. v ad x.
——————— tinctūra, ʒ ss ad iss.
———— fœtĭdi folia, ℈ i ad ij.
Hordĕi distĭchi decoctum, ℥ ij ad vj.
Hydrargȳri oxĭdum cinerĕum, gr. i ad v.
————— oxĭdum rubrum, gr. ss.
————— pilŭlæ, gr. v ad xv.
————— sub-sulphas flavus, gr. i ad v.
————— murĭas, gr. $\frac{1}{8}$ ad $\frac{1}{2}$.
————— sub-murĭas, gr. i ad v.
——————————— præcipitātus, *idem*.
————— acētis, gr. i ad vi.
————— sulphurētum nigrum, ℈ i ad ʒ i.
———————————— rubrum, gr. x ad ʒ ss.
Hyosciămi nigri herba, semen, gr. iii ad x.
——————— succus spissātus, gr. i ad v.
——————— tinctūra, ℈ i ad ʒ i.
Hyssōpi officinālis herba, ℈ i ad ʒ i.
Inŭlæ hĕlĕnii rādix, ℈ i ad ʒ i.
Ipecacuanhæ rādix, ℈ j. ad ij.
—————— vinum, gt. xxx ad ʒ j.

Ipecacuanhæ et opii pulvis, ℈ ss ad j.
Juglandis cinĕrĕæ extractum, gr. v ad ʒ ss.
Junīpĕri commūnis baccæ, ʒ ss ad i.
——————— oleum volatīle, gt. ij ad x.
——————— spirĭtus composĭtus, ʒ ij ad vi.
——— lyciæ gummi-resīna, ℈ i ad ij.
——— sabīnæ folia, gr. xv ad ℈ ij.
Kīno, gr. x ad ℈ i.
—— tinctūra, ʒ i ad iij.
Lactūcæ virōsæ succus spissātus, gr. iij ad xv.
Lauri cinnamōmi cortex, gr. v ad ℈ i.
——————— aqua destillāta, ℥ i ad iij.
——————— spirĭtus, ʒ ij ad ℥ i.
——————— tinctūra, ʒ i ad ℥ iij.
——————— oleum volatīle, gt. i ad iij.
—— camphŏræ camphŏra, gr. iij ad ℈ i.
——————— emulsio, ℥ ss ad ij.
—— nobĭlis folia, baccæ, gr. x ad ʒ ss.
—— sassăfras lignum, rādix, eorumque cortex, ℈ i ad ʒ i.
Lavandŭlæ spicæ florentes, ℈ i ad ʒ i.
————— —— tinctūra composĭta, ʒ ss ad ij.
————— oleum volatīle, gt. i ad v.
Magnēsia, gr. x ad ℈ i.
Magnēsiæ carbōnas, ℈ i ad ʒ i.

Magnēsia trochisci, ʒ i ad ij.
——— sulphas, ℥ ss ad i.
Malvæ sylvestris folia, flores, ʒ ss ad i.
Marrŭbĭi vulgāris folia, ʒ ss ad i.
Melissæ officinālis folia, gr. x ad ℈ ij.
Melŏës vesīcatorĭi pulvis, gr. ss ad i.
——————— tinctūra, gt. x ad xxx.
Menthæ virĭdis herba, gr. x ad ʒ i.
—————— aqua, ℥ i ad ij.
—————— spirĭtus, ʒ ij ad ℥ i.
—————— oleum volatĭle, gt. ij ad v.
——— pipeRītæ herba, gr. x ad ℈ ij.
——————— aqua, ℥ i ad ij.
——————— spirĭtus, ʒ ij ad ℥ i.
——————— oleum volatĭle, gt. i ad iij.
——— pūlĕgĭi herba, gr. x ad ℈ ij.
——————— aqua, ℥ i ad ij.
——————— oleum volatĭle, gt. ij ad v.
Menўanthis trĭfoliātæ folia, ʒ ss ad ℥ i.
Mimōsæ catĕchu extractum, gr. xv ad ʒ ss.
——————— electuārĭŭm, ℈ i ad ʒ i.
——————— infūsum, ℥ i ad ij.
——— catĕchu tinctūra, ʒ i ad iij.
——— nilotĭcæ gummi, ʒ i ad ij.
Momordĭcæ elaterĭi succus spissātus, gr. ss ad vi.

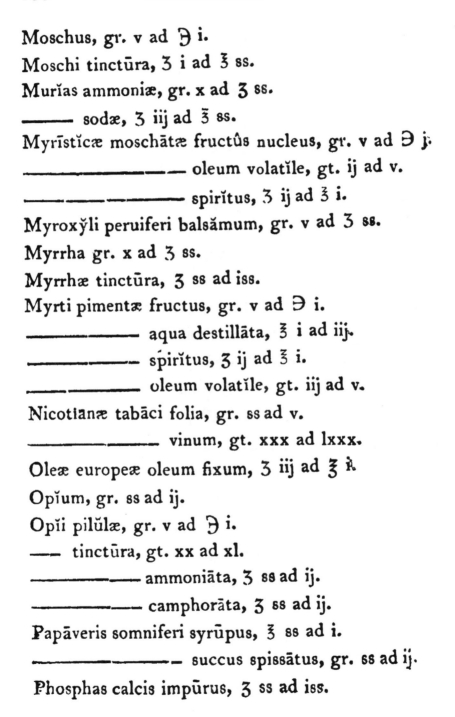

Moschus, gr. v ad ℈ i.
Moschi tinctūra, ʒ i ad ℥ ss.
Murĭas ammoniæ, gr. x ad ʒ ss.
—— sodæ, ʒ iij ad ℥ ss.
Myrīstĭcæ moschātæ fructûs nucleus, gr. v ad ℈ j.
—————— oleum volatĭle, gt. ij ad v.
—————— spirĭtus, ʒ ij ad ℥ i.
Myroxȳli peruiferi balsămum, gr. v ad ʒ ss.
Myrrha gr. x ad ʒ ss.
Myrrhæ tinctūra, ʒ ss ad iss.
Myrti pimentæ fructus, gr. v ad ℈ i.
————— aqua destillāta, ℥ i ad iij.
————— spirĭtus, ʒ ij ad ℥ i.
————— oleum volatĭle, gt. iij ad v.
Nicotianæ tabāci folia, gr. ss ad v.
—————— vinum, gt. xxx ad lxxx.
Oleæ europeæ oleum fixum, ʒ iij ad ℥ i.
Opĭum, gr. ss ad ij.
Opĭi pilŭlæ, gr. v ad ℈ i.
—— tinctūra, gt. xx ad xl.
————— ammoniāta, ʒ ss ad ij.
————— camphorāta, ʒ ss ad ij.
Papāveris somniferi syrūpus, ℥ ss ad i.
—————— succus spissātus, gr. ss ad ij.
Phosphas calcis impūrus, ʒ ss ad iss.

Physetĕris macrocephăli sevum, ʒ ss ad iss.
Phytolāccæ decandræ rādix, gr. xx ad xxx.
Pimpĭnellæ anīsi semĭna gr. xv ad ʒ ss.
———— olĕum volatĭle, gt. v ad x.
Pini balsămeæ resīna liquida, gr. v ad ʒ ss.
—— larĭcis resīna liquida, ℈ j ad ij.
—— sylvestris resīna liquida, gt. xv ad ℈ ij.
——— resīna empyreumatica, ℈ i ad ʒ i.
—— oleum volatĭle rectificātum, gt. x ad ʒ i.
Pipĕris nigri baccæ, gr. v ad ℈ i.
—— longi fructus, gr. v ad ℈ i.
Pistacíæ lentĭsci resīna, gr. v ad ʒ ss.
Plumbi acētis, gr. ss ad ij.
Polygălæ senĕgæ rādix, ℈ i ad ʒ ss.
———— decoctūm, ℥ i ad ij.
Polўgōni bistortæ rādix, gr. xv ad ʒ i.
Polўpodii filĭcis māris rādix, ʒ i ad ij.
Potassæ aqua, gt. x ad xxx.
—— acētis, ℈ i ad ʒ j.
—— super-carbōnātis aqua, ℥ vi ad ℔ ss.
—— sulphurētum, gr. v ad xv.
—— tartris, ℈ i ad ℥ ss.
—— super-tartris, ʒ i ad ℥ i.
—— sulphas, ℈ i ad ℥ ss.
—— carbōnas, gr. v ad ℈ i.

Potassæ carbōnātis aqua, ʒ ss ad i.
——— nitras, gr. v ad ʒ ss.
——— sulphas cum sulphŭre, gr. xv ad ʒ ss.
Pterŏcarpi dracōnis resīna, gr. x ad ℈ ij.
Pulvis cinnamōmi composĭtus, gr. v ad x.
—— opiātus, gr. v ad x.
Quassiæ simarūbæ cortex, ʒ ss ad i.
——— excelsæ lignum, gr. v ad ℈ i.
Quercus robŏris cortex, gr. xv ad ʒ ss.
——— cerris gallæ, gr. x ad ʒ ss.
Rhamni cathartĭci succus expressus, ℥ ss ad i.
————— syrūpus, ℥ ss ad iss.
Rhei palmāti rādix, gr. x ad ℈ ij.
———— infūsum, ℥ ss ad iss.
———— pilŭlæ composĭtæ, gr. x ad ʒ ss.
———— tinctūra, ℥ ss ad iss.
————— composĭta, ℥ ss ad iss.
———— et alŏës tinctūra, ℥ ss ad i.
———— et gentiānæ tinctūra, ℥ ss ad iss.
———— vinum, ℥ ss ad iss.
Rhŏdŏdendri chrysānthi folia, gr. v ad x.
Rhi toxicodendri folia, gr. ss ad i.
Ricĭni commūnis oleum, ℥ ss ad i.
Rosæ gallicæ petala, ℈ j ad ʒ j.
———— conserva, ʒ ij ad ℥ ss.

Rosæ gallicæ infūsum, ℥ ij ad vi.
———— syrūpus, ʒ i ad ij.
—— damascēnæ petala, ℈ i ad ʒ i.
———— aqua destillāta, ℥ i ad iij.
———— syrūpus, ʒ ij ad ℥ ss.
Rorismarīni officinālis summitates, gr. x ad ℈ ij.
———— oleum volatĭle, gt. ij ad v.
———— spirĭtus, ʒ j ad iij.
Rubĭæ tinctōrum rādix, ℈ i ad ʒ ss.
Rūtæ grăvĕŏlentis herba, gr. xv ad ℈ ij.
Sagapēnum, gr. x ad ʒ ss.
Salvĭæ officinālis folia, gr. xv ad ℈ ij.
Sambūci nigri cortex interior, gr. v ad ℈ i.
———— succus spissātus, ℥ ss ad iss.
Sapo, gr. x ad ʒ ss.
Scillæ maritĭmæ rādix recens, gr. v ad xv.
———— siccāta, gr. i ad iij.
———— syrūpus, ʒ i ad ij.
———— tinctūra, gt. x ad xx.
———— pilŭlæ, gr. x ad ℈ i.
Sinăpĕos albæ semĭna, ℥ ss ad i.
Smilācis sărsăparīllæ rādix, ℈ i ad ʒ ss.
———— sărsăparīllæ decoctum, ℥ iv ad ℔ ss.
Sodæ carbōnas, gr. x ad ʒ ss.
—— super-carbōnātis aqua, ℥ iv ad ℔ ss.

Sodæ et potassæ tartris, ʒ vj ad ℥ iss.
—— sulphas, ℥ ss ad iss.
—— phosphas, ℥ ss ad iss.
—— sub-boras, gr. x ad ʒ ss.
Spigēliæ marilandĭcæ rādix, ʒ ss ad ℈ ij.
Spirĭtus ætheris nitrōsi, ʒ ss ad j.
Stanni pulvis et limatūra, ʒ i ad ij.
—— amalgamatis pulvis, ℈ i ad ij.
Styrăcis officinālis balsămum, gr. x ad ʒ ss.
—— benzoïn balsămum, gr. x ad ʒ ss.
—— tinctūra composĭta, ʒ ss ad i.
Succĭni oleum purissimum, gt. x ad xx.
Sulphur sublimātum lotum, ℈ i ad ʒ i.
Tamărindi indícæ fructus, ℥ ss ad iss.
—— infūsum cum cassia senna, ℥ ij ad iv.
Tanacēti vulgāris flores, folia, ʒ ss ad i.
Toluifĕræ balsămi balsămum, gr. xv ad ℈ ij.
—— syrūpus, ʒ i ad iij.
—— tinctūra, ʒ ss ad ij.
Tormentĭllæ erectæ rādix, ℈ i ad ij.
Valeriānæ officinālis rādix, ℈ i ad ʒ i.
Verātri albi rādix, gr. v ad ℈ i.
—— tinctūra, gt. v ad x.
Viŏlæ odorātæ syrūpus, ʒ i ad ij.
Zinci oxĭdum, gr. iij ad x.
—— sulphas, gr. vi ad ʒ ss.

Note. These are in general the doses for adults from twenty to sixty, but they may be diminished for children and people past the prime of life, nearly in the following proportions:

	Ages.	Proportionate doses.
Months	2	$\frac{1}{15}$
	7	$\frac{1}{12}$
	14	$\frac{1}{8}$
	28	$\frac{1}{5}$
Years	3	$\frac{1}{4}$
	5	$\frac{1}{3}$
	7	$\frac{1}{2}$
	14	$\frac{2}{3}$
	63	$\frac{11}{12}$
	77	$\frac{5}{6}$
	100	$\frac{4}{6}$

It may also be observed, that sixty drops of water, one hundred of diluted alcohol, or an hundred and twenty of alcohol, are equal to a drachm by measure.

TABLE

OF

ANCIENT NAMES WITH THEIR SYNONIMES

IN THE

NOMENCLATURE OF THIS WORK.

Ancient names.	*Systematic names.*
Abrotanum.	Artemisia abrotanum.
Absinthium.	——— absinthium.
Acetosella.	Oxalis acetosella.
Acetum vini.	Acidum acetosum.
—— *distillatum.*	Acidum acetosum destillatum.
—— *scilliticum.*	Acetum scillæ maritimæ.
Acidum vitriolicum.	Acidum sulphuricum.
Aconitum.	Aconitum neomontanum.
Ærugo.	Sub-acetis cupri.
Æther vitrioli.	Æther sulphuricus.
Æthiops martialis.	Carbonas ferri.
——— *mineralis.*	Sulphuretum hydrargyri nigrum.
Alkali causticum	Potassa.
—— *fixum fossile.*	Carbonas sodæ impurus.
—— —— *vegetabile.*	——— potassæ impurus.
—— *volatile.*	——— ammoniæ.
Aloë.	Aloë perfoliata.

Ancient names.	*Systematic names.*
Althæa.	Althæa officinalis.
Alumen.	Super-sulphas aluminæ et potassæ.
——— *ustum.*	Super-sulphas aluminæ et potassæ exsiccatus.
Ammonia.	Carbonas ammoniæ.
Aconitum.	Aconitum neomontanum.
Amygdala dulcis.	Amygdalus communis.
Anethum.	Anethum graveolens.
Angelica.	Angelica archangelica.
Anisum.	Pimpinella anisum.
Antimonium.	Sulphuretum antimonii.
——— *præparatum.*	——— ——— præparatum.
——— *tartarizatum.*	Tartris antimonii.
Aqua aëris fixi.	Aqua acidi carbonici.
—— *ammoniæ.*	—— carbonatis ammoniæ.
—— ——— *causticæ.*	—— ammoniæ.
—— *cassiæ ligneæ.*	—— lauri cassiæ.
—— *cinnamomi simplex.*	—— --- cinnamomi.
—— ——— *spirituosa.*	Spiritus lauri cinnamomi.
—— *ferri aërati.*	Aqua super-carbonatis ferri.
—— *fortis.*	Acidum nitrosum.
—— *kali præparati.*	Aqua carbonatis potassæ.
—— — *puri.* } —— *lixivia caustica.* }	—— potassæ.
—— *pimentæ.*	—— myrti pimentæ.
—— *pulegii.*	—— menthæ pulegii.
—— *rosarum.*	—— rosæ damascenæ.
—— *styptica.*	Solutio sulphatis cupri composita.
Argentum vivum.	Hydrargyrus.
——— ——— *purificatum.*	——— ——— purificatus.

Ancient names.	*Systematic names.*
Arnica.	Arnica montana.
Arsenicum. / ——— *album.*	Oxidum arsenici.
Assa fœtida.	Gummi-resina ferulæ assæ fœtidæ.
Aurantium hispalense.	Citrus aurantium.
Avena.	Avena sativa.
Axungia porcina.	Adeps suis scrofæ.
Balsamum anodynum.	Tinctura saponis et opii
——— *canadense.*	Resina liquida pini balsameæ.
——— *copaibæ.*	Resina liquida copaiferæ officinalis.
——— *gileadense.*	Resina liquida amyridis gileadensis.
——— *peruvianum.*	Balsamum myroxyli peruiferi.
——— *saponaceum.*	Tinctura saponis.
——— *sulphuris.*	Oleum sulphuratum.
——— *tolutanum.*	Balsamum toluiferæ balsami.
——— *traumaticum.*	Tinctura benzoin composita.
Bardana.	Arctium lappa.
Barilla.	Carbonas sodæ impurus.
Barytes.	——— barytæ.
——— *muriatus.*	Murias barytæ.
Belladonna.	Atropa belladonna.
Benzoïnum.	Balsamum styracis benzoïn.
Bistorta.	Polygonum bistorta.
Borax.	Sub-boras sodæ.
Butyrum antimonii.	Murias antimonii.
Cajeputa.	Melaleuca leucadendron.
Calamus aromaticus.	Acorus calamus.
Calomelanos. / *Calomelas.*	Sub-murias hydrargyri

Ancient names.	*Systematic names.*
Calx hydrargyri alba.	Sub-murias hydrargyri et ammoniæ.
— *viva.*	Calx.
Camphora.	Laurus camphora.
Cancrorum oculi. }	Carbonas calcis durior.
——— *lapilli.* }	
Cantharis.	Meloë vesicatorius.
Cardamomum minus.	Amomum repens.
Carduus benedictus.	Centaurea benedicta.
Carica.	Fructus ficûs caricæ.
Caryophyllus aromaticus.	Eugenia caryophyllata.
Carvi.	Carum carui.
Cascarilla.	Croton eleutheria.
Cassia.	Laurus cassia.
Catechu.	Extractum mimosæ catechu.
Causticum commune acerrimum.	Potassa.
——— ——— *mitius.*	Potassa cum calce.
——— *lunare.*	Nitras argenti.
Centaurium minus.	Chironia centaurium.
Ceratum epuloticum. }	Ceratum carbonatis zinci impuri.
——— *lapidis calaminaris.* }	
——— *turneri.* }	
Cerussa.	Oxidum plumbi album.
Chamæmelum.	Anthemis nobilis.
Cicuta.	Conium maculatum.
Cineres clavellati.	Carbonas potassæ impurus.
Cinnabaris factitia.	Sulphuretum hydrargyri rubrum.
Cinnamomum.	Laurus cinnamomum.
Coccinella.	Coccus cacti.
Cochlearia.	Cochlearia officinalis.
Colchicum.	Colchicum autumnale.
Colcothar vitrioli.	Oxidum ferri rubrum.

Ancient names.	*Systematic names.*
Colocynthis.	Cucumis colocynthis.
Confectio cardiaca.	Electuarium aromaticum.
Confectio japonica.	Electuarium catechu.
Conserva corticis aurantii.	Conserva citri aurantii.
——— rosarum.	——— rosæ damascenæ.
Contrayerva.	Dorstenia contrajerva.
Cornu cervi.	Cornu cervi elaphi.
—— —— ustum.	Phosphas calcis impurus.
Coriandrum.	Coriandrum sativum.
Cortex peruvianus.	Cortex cinchonæ officinalis.
Cremor tartari.	Super-tartris potassæ.
Creta alba.	Carbonas calcis mollior.
—— præparata.	——— —— præparatus.
Crocus antimonii.	Oxidum antimonii cum sulphure per nitratem potassæ.
—— anglicus.	Crocus sativus.
—— martis.	Oxidum ferri rubrum.
—— metallorum.	——— antimonii cum sulphure per nitratem potassæ.
Crystalli tartari.	Super-tartris potassæ.
Cuminum.	Cuminum cyminum.
Cuprum ammoniacum.	Ammoniaretum cupri.
——— vitriolatum.	Sulphas cupri.
Curcuma.	Curcuma longa.
Cynosbatus.	Rosa canina.
Daucus sylvestris.	Daucus carota.
Decoctum lignorum.	Decoctum guajaci compositum.
Dens leonis.	Leontodon taraxacum.
Digitalis.	Digitalis purpurea.
Dolichos.	Dolichos pruriens.
Dulcamara.	Solanum dulcamara.

Ancient names.	*Systematic names.*
Elaterium.	Succus spissatus momordicæ elaterii.
Electuarium lenitivum.	Electuarium cassiæ sennæ.
——— *thebaicum.*	——— opiatum.
Eleutheria.	Croton eleutheria.
Elixir asthmaticum.	Tinctura opii camphorata.
—— *camphoræ.*	——— camphoræ.
—— *paregoricum.*	——— opii camphorata.
—— *proprietatis.*	——— aloës et myrrhæ.
—— *sacrum.*	——— rhei et aloës.
—— *salutis.*	——— sennæ composita.
—— *stomachicum.*	——— gentianæ composita.
—— *vitrioli.*	Acidum sulphuricum aromaticum.
Emplastrum adhesivum.	Emplastrum resinosum.
——— *antisthericum.*	——— assæ fœtidæ.
——— *cantharidum.*	——— meloës vesicatorii.
——— *cereum.*	——— simplex.
——— *cerussæ.*	——— oxidi plumbi albi.
——— *cæruleum.*	——— hydrargyri.
——— *commune.*	——— oxidi plumbi semivitrei.
——— ——— *cum gummis.*	——— gummosum.
——— *diachyli.*	——— oxidi plumbi semivitrei.
——— *epispasticum.*	——— meloës vesicatorii.
——— *mercuriale.*	——— hydrargyri.
——— *picis burgundicæ.*	——— resinosum compositum.
——— *roborans.*	——— oxidi ferri rubri.
——— *vesicatorium.*	——— meloës vesicatorii.
Emulsio communis.	Emulsio amygdali communis.

Ancient names.	*Systematic names.*
Ens veneris.	Murias ammoniæ et ferri.
Enula campana.	Inula helenium.
Essentia antimonialis.	Vinum tartritis antimonii.
Extractum capitum papaveris albi.	Extractum papaveris somniferi.
——— *cicutæ.*	Succus spissatus conii maculati.
——— *chamæmeli.*	Extractum anthemidis nobilis.
——— *jalapæ.*	-------- convolvuli jalapæ.
——— *ligni campechensis.*	-------- hæmatoxyli campechensis.
——— *nucis butyraceæ.*	——— juglandis cinereæ.
——— *sennæ.*	-------- cassiæ ſennæ.
Ferri rubigo.	Carbonas ferri.
—— *squamæ.*	Oxidum ferri nigrum.
—— —— *purificatæ.*	Oxidum ferri nigrum purificatum.
Ferrum ammoniacale.	Murias ammoniæ et ferri.
Filix mas.	Polypodium filix mas.
Flores benzoïni.	Acidum benzoicum.
—— *martiales.*	Murias ammoniæ et ferri.
—— *zinci.*	Oxidum zinci.
Fœniculum dulce.	Anethum fœniculum.
Galbanum.	Bubon galbanum.
Gentiana.	Gentiana lutea.
Geoffræa.	Geoffræa inermis.
Glycirrhiza.	Glycirrhiza glabra.
Gratiola.	Gratiola officinalis.
Guaiacum.	Guajacum officinale.
Gummi arabicum.	Gummi mimosæ niloticæ.
Helleboraster.	Helleborus fœtidus.
Helleborus albus.	Veratrum album.

Ancient names.	*Systematic names.*
Hepar sulphuris.	Sulphuretum potassæ.
Hippocastanum.	Æsculus hippocastanum.
Hordeum.	Hordeum distichon.
Hydrargyrus muriatus corrosivus.	Murias hydrargyri.
——— ——— *mitis.*	Sub-murias hydrargyri.
Hyoscyamus.	Hyoscyamus niger.
Hyssopus.	Hyssopus officinalis.
Infusum amarum.	Infusum gentianæ compositum.
——— *japonicum.*	——— mimosæ catechu.
Iris.	Iris pseudacorus.
Jalapa.	Convolvulus jalapa.
Kali purum.	Potassa.
––– *præparatum.*	Carbonas potassæ.
Lac ammoniaci.	Emulsio ammoniaci.
Lactuca.	Lactuca sativa.
Lapis calaminaris.	Carbonas zinci impurus.
—— *infernalis.* / —— *septicus.*	Potassa.
Laudanum liquidum.	Tinctura opii.
Lavandula.	Lavandula spica.
Lignum campechense.	Hæmatoxylum campechianum.
Limon.	Citrus medica.
Linimentum saponaceum.	Tinctura saponis.
——— *volatile.*	Oleum ammoniatum.
Liquor alkali vegetabilis mitissimi.	Aqua super-carbonatis potassæ.
Linum.	Linum usitatissimum.
Lithargyrus. / ——— *auri.*	Oxidum plumbi semivitreum.
Lixivium causticum.	Aqua potassæ.

Ancient names.	Systematic names.
Lupulus.	Humulus lupulus.
Magnesia. ——— *alba.*	Carbonas magnesiæ.
——— *calcinata.* ——— *usta.*	Magnesia.
——— *vitriolata.*	Sulphas magnesiæ.
Majorana.	Origanum majorana.
Malva.	Malva sylvestris.
Manna.	Succus concretus fraxini orni.
Marmor album.	Carbonas calcis durior.
Mars.	Ferrum.
Marrubium.	Marrubium vulgare.
Masticbe.	Pistacia lentiscus.
Melampodium.	Helleborus niger.
Melissa.	Melissa officinalis.
Mentha piperitis.	Mentha piperita.
——— *sativa.*	——— viridis.
Mercurius.	Hydrargyrus.
——— *calcinatus.*	Oxidum hydrargyri cinereum
——— *corrosivus sublimatus.*	Murias hydrargyri.
——— *dulcis.*	Sub-murias hydrargyri.
——— *emeticus flavus.*	Sub-sulphas hydrargyri flavus.
——— *præcipitatus albus.*	Sub-murias hydrargyri et ammoniæ.
——— ——— *ruber.*	Oxidum hydrargyri rubrum per acidum nitricum.
Mezereum.	Daphne mezereum.
Minium.	Oxidum plumbi rubrum.
Mucilago arabici gummi.	Mucilago mimosæ niloticæ
Muria.	Murias sodæ.
Natron præparatum.	Carbonas sodæ.

Ancient names.	*Systematic names.*
Nicotiana.	Nicotiana tabacum.
Nitrum.	Nitras potassæ.
Nux moschata.	Nucleus fructûs myristicæ moschatæ.
Oculi cancrorum.	Carbonas calcis durior.
Olea essentialia.	Olea volatilia.
Olibanum.	Gummi-resina juniperi lyciæ.
Oliva.	Olea europæa.
Oleum essentiale juniperi.	Oleum volatile juniperi communis.
—— *succini rectificatum.*	—— succini purissimum.
—— *tartari per deliquium.*	Aqua carbonatis potassæ.
—— *terebinthinæ.*	Oleum volatile pini laricis.
—— ——— *rectificatum.*	—— ——— —— purissimum.
—— *vitrioli.*	Acidum sulphuricum.
Petroleum barbadense.	Bitumen petroleum.
Petroselinum.	Apium petroselinum.
Pilulæ cochiæ.	Pilulæ aloës et colocynthidis.
—— *cupri.*	----- ammoniareti cupri.
—— *fœtidæ.*	----- assæ fœtidæ compositæ.
—— *mercuriales.*	----- hydrargyri.
—— *rufi.*	----- aloës et myrrhæ.
—— *saponaceæ.*	----- aloëticæ.
—— *stomachicæ.*	----- rhei compositæ.
—— *thebaicæ.*	----- opiatæ.
Pimento.	Myrtus pimenta.
Piper indicum.	Capsicum annuum.
—— *jamaicense.*	Myrtus pimenta.
Pix burgundica.	Resina sponte concreta pini abietis.

Ancient names.	*Systematic names.*
Pix liquida.	Resina empyreumatica pini sylvestris.
Pranus gallica.	Prunus domestica.
Potio cretacea.	Mistura carbonatis calcis.
Pulegium.	Mentha pulegium.
Pulvis antimonialis.	Oxidum antimonii cum phosphate calcis.
---- *aromaticus.*	Pulvis cinnamomi compositus.
---- *asarabaccæ.*	----- asari compositus.
---- *cretaceus.*	----- carbonatis calcis compositus.
---- *doveri.*	----- ipecacuanhæ et opii.
---- *stypticus helvetii.*	----- super-sulphatis aluminæ et potassæ compositus.
Pyrethrum.	Anthemis pyrethrum.
Quassia.	Quassia excelsa.
Quercus.	Quercus robur.
Raphanus rusticanus.	Cochlearia armoracia.
Resina alba.	Resina pini.
Rhabarbarum. / *Rheum.*	Rheum palmatum.
Rhododendron.	Rhododendron crysanthum
Ricinus.	Ricinus communis.
Rob sambuci.	Succus spissatus sambuci nigræ
Rosa centifolia. / --- *pallida.*	Rosa damascena.
--- *rubra.*	--- gallica.
Rosmarinus.	Rosmarinus officinalis.
Rubia.	Rubia tinctorum.
Rubigo ferri.	Carbonas ferri,
Ruta.	Ruta graveolens.

Ancient names.	*Systematic names.*
Sabina.	Juniperus sabina.
Saccharum saturni.	Acetis plumbi.
Sal absynthii.	Carbonas potassæ purissimus.
— *alkalinus fixus fossilis purificatus.*	------- sodæ.
— *alkalinus fixus vegetabilis purificatus.*	Carbonas potassæ.
— *ammoniacus.*	Murias ammoniæ.
— *catharticus amarus.*	Sulphas magnesiæ.
— *cornu cervi.*	Carbonas ammoniæ.
— *diureticus.*	Acetis potassæ.
— *epsomiensis.*	Sulphas magnesiæ.
— *marinus.*	Murias sodæ.
— *martis.*	Sulphas ferri.
— *glauberi.*	——— sodæ.
— *nitri.*	Nitras potassæ.
— *polychrestus.*	Sulphas potassæ cum sulphure.
— *rupellensis.*	Tartris potassæ et sodæ.
— *succini.*	Acidum succinicum.
— *tartari.*	Carbonas potassæ purissimus.
— *vitrioli.*	Sulphas zinci.
— *volatile salis ammoniaci.*	Carbonas ammoniæ.
Salvia.	Salvia officinalis.
Sambucus.	Sambucus nigra.
Sanguis draconis.	Resina pterocarpi dracontis.
Santalum rubrum.	Lignum pterocarpi santalini.
Santonicum.	Artemisia santonica.
Sapo albus hispanus.	Sapo.
Sarsaparilla.	Smilax sarsaparilla.
Sassafras.	Laurus sassafras.
Saturnus.	Plumbum.
Scammonium.	Gummi-resina convolvuli scammoniæ.

Ancient names.	*Systematic names.*
Scilla.	Scilla maritima.
Sel de seignette.	Tartris potassæ et sodæ.
Seneka.	Polygala senega.
Senna.	Cassia senna.
Serpentaria virginiana.	Aristolochia serpentaria.
Sevum ovillum.	Adeps ovis arietis.
Simarouba.	Quassia simaruba.
Sinapi album.	Sinapis alba.
Soda.	Carbonas sodæ.
— *muriata.*	Murias sodæ.
Solanum lethale.	Atropa belladonna.
Solutio terræ ponderosæ salitæ.	Solutio muriatis barytæ.
Species aromaticæ.	Pulvis cinnamomi compositus.
Spermaceti.	Sevum physeteris macrocephali.
Spigelia.	Spigelia marilandica.
Spina cervina.	Rhamnus catharticus.
Spiritus ammoniæ.	Alcohol ammoniatum.
——— *ammoniæ aromaticus.*	——— ammoniatum aromaticum.
——— ——— *fœtidus.*	——— ——— fœtidum.
——— *carvi.*	Spiritus cari carui.
——— *cinnamomi.*	------ lauri cinnamomi.
——— *cornu cervi.*	Aqua carbonatis ammoniæ.
——— *mindereri.*	Aqua acetitis ammoniæ.
——— *nucis moschatæ.*	Spiritus myristicæ moschatæ.
——— *nitri dulcis.*	——— ætheris nitrosi.
——— — *fortis.*	Acidum nitricum.
——— *pimentæ.*	Spiritus myrti pimentæ.
——— *salis marini.*	Acidum muriaticum.
——— *salis ammoniaci.*	Aqua carbonatis ammoniæ.
——— — ——— *vinosus.*	Alcohol ammoniatum.
——— *vinosus camphoratus.*	Tinctura camphoræ.

Ancient names.	*Systematic names.*
Spiritus vinosus rectificatus.	Alcohol.
——— ——— *tenuior.*	——— dilutum.
——— *vitrioli dulcis.*	Æther sulphuricus cum alcohole.
——— *vitrioli fortis.*	Acidum sulphuricum.
——— ——— *tenuis.*	——— ——— dilutum.
Spongia.	Spongia officinalis.
Staphisagria.	Delphinium staphisagria.
Stibium.	Sulphuretum antimonii.
Stramonium.	Datura stramonium.
Styrax.	Styrax officinalis.
Succi ad scorbuticos.	Succus cochleariæ officinalis compositus.
Sulphur antimonii præcipitatum vel auratum.	Sulphuretum antimonii præcipitatum.
Sulphuris flores.	Sulphur sublimatum.
Syrupus balsami tolutani.	Syrupus toluiferæ balsami.
——— *corticis aurantii.*	——— citri aurantii.
——— *ex althæa.*	——— althææ officinalis.
——— *limonum.*	——— citri medicæ.
——— *papaveris albi.*	——— papaveris somniferi.
——— *rosæ solutivus.*	——— rosæ damascenæ.
——— *spinæ cervinæ.*	——— rhamni cathartici.
——— *zingiberis.*	——— amomi zingiberis.
Tamarindus.	Tamarindus indica.
Tanacetum.	Tanacetum vulgare.
Tartarum regeneratum.	Acetis potassæ.
——— *solubile.*	Tartris potassæ.
——— *vitriolatum.*	Sulphas potassæ.
Tartarus emeticus.	Tartris antimonii.
——— *crudus.*	Super-tartris potassæ impurus.
——— *purificatus.*	Super-tartris potassæ.

Ancient names.	*Systematic names.*
Terebinthina veneta.	Resina liquida pini laricis.
Terra japonica.	Extractum mimosæ catechu.
—— *ponderosa.*	Carbonas barytæ.
—— —— *vitriolata.*	Sulphas barytæ.
Tinctura alexipharmica.	Tinctura cinchonæ composita.
—— *amara.*	—— gentianæ composita.
—— *aromatica.*	—— lauri cinnamomi composita.
—— *balsami tolutani.*	—— toluiferæ balsami.
—— *cantharidum.*	—— meloës vesicatorii.
—— *cardamomi.*	—— amomi repentis.
—— *cascarillæ.*	—— eleutheriæ.
—— *cinnamomi.*	—— lauri cinnamomi.
—— *corticis peruviani.*	—— cinchonæ officinalis.
—— —— —— *composita.*	—— —— composita.
—— *ferri.*	—— muriatis ferri.
—— *fœtida.*	—— ferulæ assæ fœtidæ.
—— *guaiaci volatilis.*	—— guajaci ammoniata.
—— *hellebori albi.*	—— veratri albi.
—— *ipecacuanhæ.*	Vinum ipecacuanhæ.
—— *jalapæ.*	Tinctura convolvuli jalapæ.
—— *japonica.*	—— mimosæ catechu.
—— *martis.*	—— muriatis ammoniæ et ferri.
—— —— *salita.*	—— muriatis ferri.
—— *rhei.*	—— rhei palmati.
—— — *amara.*	—— --- et gentianæ.
—— *sacra.*	Vinum aloës socotorinæ.
—— *serpentariæ.*	Tinctura aristolochiæ serpentariæ.
—— *thebaica.*	—— opii.
Tormentilla.	Tormentilla erecta.

Ancient names.	*Systematic names.*
Toxicodendron.	Rhus toxicodendron.
Tragacantha.	Gummi astragali tragacanthæ.
Trifolium palustre.	Menyanthes trifoliata.
Triticum.	Triticum æstivum.
Trochisci cretæ.	Trochisci carbonatis calcis.
——— *glycirrhizæ compositi.*	——— glycirrhizæ cum opio.
Turpethum minerale.	Sub-sulphas hydrargyri flavus.
Tussilago.	Tussilago farfara.
Tutia.	Oxidum zinci impurum.
—— *præparata.*	——— ——— ——— præparatum.
Unguentum æruginis.	Unguentum sub-acetitis cupri.
——— *album.* } ——— *è cerussa.* }	——— oxidi plumbi albi.
——— *cæruleum.*	——— hydrargyri.
——— *basilici flavi.*	——— resinosum.
——— *calcis hydrargyri albi.*	——— sub-muriatis hydrargyri et ammoniæ.
——— *citrinum.*	——— nitratis hydrargyri.
——— *epispasticum fortius.*	——— pulveris meloës vesicatorii.
——— ——— *mitius.*	——— infusi meloës vesicatorii.
——— *è mercurio præcipitato.*	——— sub-muriatis hydrargyri et ammoniæ.
——— *mercuriale fortius.*	——— hydrargyri.
——— ——— *mitius.*	——— ——— mitius.
——— *oxygenatum.*	——— acidi nitrosi.
——— *saturninum.*	——— acetitis plumbi.
——— *tutiæ.*	——— oxidi zinci impuri.

Ancient names.	*Systematic names.*
Unguentum zinci.	Unguentum oxidi zinci.
Uva passa.	Fructus siccatus vitis viniferi.
--- *ursi.*	Arbutus uva ursi.
Valeriana sylvestris.	Valeriana officinalis.
Vinum amarum.	Vinum gentianæ compositum.
——- *antimoniale.*	——— tartritis antimonii.
——- *chalybeatum.*	——— ferri.
Viola.	Viola odorata.
Vitriolum album.	Sulphas zinci.
——— *cæruleum.* } ——— *romanum.* }	------ cupri.
——— *viride.*	------ ferri.
Vitrum antimonii.	Oxidum antimonii cum sulphure vitrificatum.
——— ——— *ceratum.*	Oxidum antimonii vitrificatum cum cera.
Zedoaria.	Amomum zedoaria.
Zincum vitriolatum.	Sulphas zinci.
Zingiber.	Amomum zingiber.

NOTE. This table is designed to present the names, in use among us only. We have not inserted those, adopted some years ago by the London and Edinburgh colleges, because the greater part of them have never been employed by our physicians.

TABLE

OF

SYSTEMATIC NAMES USED IN THIS WORK

WITH THEIR

SYNONIMES IN THE ANCIENT NOMENCLATURE.

Systematic names.	Ancient names.
Acetis plumbi.	*Saccharum saturni.*
—— potassæ.	*Tartarum regeneratum.* *Sal diureticus.*
Acetum scillæ maritimæ.	*Acetum scilliticum.*
Acidum acetosum.	—— *vini.*
—— —— destillatum.	—— *distillatum.*
—— benzoïcum.	*Flores benzoïni.*
—— muriaticum.	*Spiritus salis marini.*
—— nitricum.	—— *nitri fortis.*
—— nitrosum.	*Aqua fortis.*
—— succinicum.	*Sal succini.*
—— sulphuricum.	*Acidum vitriolicum.* *Oleum vitrioli.* *Spiritus vitrioli fortis.*
—— sulphuricum aromaticum.	*Elixir vitrioli.*
—— —— dilutum.	*Spiritus vitrioli tenuis.*
Aconitum neomontanum.	*Aconitum.*

Systematic names.	Ancient names.
Acorus calamus.	*Calamus aromaticus*
Æsculus hippocastanum.	*Hippocastanum.*
Æther sulphuricus.	*Æther vitrioli.*
—— —— cum alcohole.	*Spiritus ætheris vitriolici.* / —— *vitrioli dulcis.*
Alcohol.	—— *vinosus rectificatus.*
—— ammoniatum.	—— *ammoniæ.* / —— *salis ammoniaci vinosus.*
—— —— aromaticum.	—— — —— *aromaticus.*
—— —— fœtidum.	—— — —— *fœtidus.*
—— dilutum.	—— *vinosus tenuior.*
Aloë perfoliata.	*Aloë.*
Althæa officinalis.	*Althæa.*
Ammoniaretum cupri.	*Cuprum ammoniacum.*
Amomum repens.	*Cardamomum minus.*
—— zedoaria.	*Zedoaria.*
—— zingiber.	*Zingiber.*
Amygdalus communis.	*Amygdala dulcis.*
Amyridis gileadensis resina liquida.	*Balsamum gileadense.*
Anethum fœniculum.	*Fœniculum dulce.*
Anethum graveolens.	*Anethum.*
Angelica archangelica.	*Angelica.*
Anthemis nobilis.	*Chamæmelum.*
—— pyrethrum.	*Pyrethrum.*
Apium petroselinum.	*Petroselinum.*
Aqua acetitis ammoniæ.	*Spiritus mindereri.*
—— acidi carbonici.	*Aqua aëris fixi.*
—— ammoniæ.	—— *ammoniæ causticæ.*
—— carbonatis ammoniæ.	—— *ammoniæ.* / *Spiritus cornu cervi.* / —— *salis ammoniaci.*

Systematic names.	Ancient names.
Aqua carbonatis potassæ.	*Aqua kali præparati.*
	Oleum tartari per deliquium.
—— super-carbonatis ferri.	*Aqua ferri aërati.*
—— lauri cassiæ.	—— *cassiæ ligneæ.*
—— --- cinnamomi.	—— *cinnamomi simplex.*
—— menthæ pulegii.	—— *pulegii.*
—— myrti pimentæ.	—— *pimentæ.*
—— potassæ.	*Lixivium causticum.*
	Aqua lixivia caustica.
	—— *kali puri.*
—— rosæ damascenæ.	—— *rosarum.*
—— super-carbonatis potassæ.	*Liquor alkali vegetabilis mitissimi.*
Arbutus uva ursi.	*Uva ursi.*
Arctium lappa.	*Bardana.*
Aristolochia serpentaria.	*Serpentaria virginiana.*
Arnica montana.	*Arnica.*
Artemisia abrotanum.	*Abrotanum.*
——— absinthium.	*Absinthium.*
——— santonica.	*Santonicum.*
Astragali tragacanthæ gummi.	*Tragacantha.*
Atropa belladonna.	*Belladonna.*
	Solanum lethale.
Avena sativa.	*Avena.*
Bitumen petroleum.	*Petroleum barbadense.*
Bubon galbanum.	*Galbanum.*
Calx.	*Calx viva.*
Capsicum annuum.	*Piper indicum.*
Carbonas ammoniæ.	*Ammonia.*
	Sal cornu cervi.
	— *volatile salis ammoniaci.*
	Alkali volatile.

Systematic names.	Ancient names.
Carbonas barytæ.	*Barytes.* *Terra ponderosa.*
——— calcis mollior.	*Creta alba.*
——— ——— durior.	*Marmor album.* *Cancrorum oculi vel lapilli.*
——— ——— præparatus.	*Creta præparata.*
——— ferri.	*Rubigo ferri.* *Æthiops martialis.*
——— magnesiæ.	*Magnesia.* ——— *alba.*
——— potassæ.	*Sal alkalinus fixus vegetabilis purificatus.* *Kali præparatum.*
——— ——— impurus.	*Alkali fixum vegitabile.* *Cineres clavellati.*
——— ——— purissimus.	*Sal tartari.* -- *absynthii.*
——— sodæ.	*Natron præparatum.* *Soda.* *Sal alkalinus fixus fossilis purificatus.*
——— ——— impurus.	*Barilla.* *Alkali fixum fossile.*
——— zinci impurus.	*Lapis calaminaris.*
——— ——— ——— præparatus.	——— ——— *præparatus.*
Carum carui.	*Carvi.*
Cassia senna.	*Senna.*
Centaurea benedicta.	*Carduus benedictus.*
Ceratum carbonatis zinci impuri.	*Ceratum epuloticum.* ——— *lapidis calaminaris.* ——— *turneri.*
Cervi elaphi cornu.	*Cornu cervi.*

Systematic names.	Ancient names.
Chironia centaurium.	*Centaurium minus.*
Cinchonæ officinalis cortex.	*Cortex peruvianus.*
Citrus aurantium.	*Aurantium hispalense.*
Citrus medica.	*Limon.*
Coccus cacti.	*Coccinella.*
Cochlearia armoracia.	*Raphanus rusticanus.*
——— officinalis.	*Cochlearia.*
Colchicum autumnale.	*Colchicum.*
Conium maculatum.	*Cicuta.*
Conserva citri aurantii.	*Conserva corticis aurantii.*
——— rosæ gallicæ.	——— *rosarum.*
Convolvuli scammoniæ gummi-resina.	*Scammonium.*
Convolvulus jalapa.	*Jalapa.*
Copaiferæ officinalis resina liquida.	*Balsamum copaibæ.*
Coriandrum sativum.	*Coriandrum.*
Crocus sativus.	*Crocus anglicus.*
Croton eleutheria.	{ *Cascarilla.* { *Eleutheria.*
Cucumis colocynthis.	*Colocynthis.*
Cuminum cyminum.	*Cuminum.*
Curcuma longa.	*Curcuma.*
Daphne mezereum.	*Mezereum.*
Datura stramonium.	*Stramonium.*
Daucus carota.	*Daucus sylvestris.*
Decoctum guajaci compositum.	*Decoctum lignorum.*
Delphinium staphisagria.	*Staphisagria.*
Digitalis purpurea.	*Digitalis.*
Dolichos pruriens.	*Dolichos.*
Dorstenia contrajerva.	*Contrayerva.*

Systematic names.	Ancient names.
Electuarium aromaticum.	*Confectio cardiaca.*
——— cassiæ sennæ.	*Electuarium lenitivum.*
——— catechu.	*Confectio japonica.*
——— opiatum.	*Electuarium thebaicum.*
Emplastrum assæ fœtidæ.	*Emplastrum antisthericum.*
——— hydrargyri.	{ ——— *mercuriale.*
	{ ——— *cæruleum.*
——— resinosum.	——— *adhesivum.*
——— resinosum compositum.	——— *picis burgundicæ.*
——— gummosum.	——— *commune cum gummis.*
——— meloës vesicatorii.	{ ——— *epispasticum.*
	{ ——— *cantharidum.*
	{ ——— *vesicatorium.*
——— oxidi ferri rubri.	——— *roborans.*
——— oxidi plumbi albi.	——— *cerussæ.*
——— ——— ——— semivitrei.	{ ——— *commune.*
	{ ——— *diachyli.*
——— simplex.	——— *cereum.*
Emulsio ammoniaci.	*Lac ammoniaci.*
——— amygdali communis.	*Emulsio communis.*
Eugenia caryophyllata.	*Caryophyllus aromaticus.*
Extractum anthemidis nobilis.	*Extractum chamæmeli.*
——— cassiæ fennæ.	——— *sennæ.*
——— convolvuli jalapæ.	——— *jalapæ.*
——— hæmatoxyli campechensis.	——— *ligni campechensis.*
——— juglandis cinereæ.	——— *nucis butyraceæ.*
——— papaveris somniferi.	——— *capitum papaveris albi*
Ferri oxidum nigrum purificatum.	*Ferri squamæ purificatæ.*

Systematic names.	Ancient names.
Ferrum.	*Mars.*
Ferulæ assæ fœtidæ gummi-resina.	*Assa fœtida.*
Ficûs caricæ fructus.	*Carica.*
Fraxini orni succus concretus.	*Manna.*
Gentiana lutea.	*Gentiana.*
Geoffræa inermis.	*Geoffræa.*
Glycirrhiza glabra.	*Glycirrhiza.*
Gratiola officinalis.	*Gratiola.*
Guajacum officinale.	*Guaiacum.*
Hæmatoxylum campechianum.	*Lignum campechense.*
Helleborus fœtidus.	*Helleboraster.*
——— niger.	*Melampodium.*
Hordeum distichon.	*Hordeum.*
Humulus lupulus.	*Lupulus.*
Hydrargyrus.	{ *Argentum vivum.* *Mercurius.*
——— purificatus.	*Argentum vivum purificatum.*
Hyoscyamus niger.	*Hyoscyamus.*
Hyssopus officinalis.	*Hyssopus.*
Infusum gentianæ compositum.	*Infusum amarum.*
——— mimosæ catechu.	——— *japonicum.*
Inula helenium.	*Enula campana.*
Iris pseudacorus.	*Iris.*
Juniperi lyciæ gummi-resina.	*Olibanum.*
Juniperus sabina.	*Sabina.*
Lactnca sativa.	*Lactuca.*
Laurus camphora.	*Camphora.*

Systematic names.	Ancient names.
Laurus cassia.	*Cassia.*
——— cinnamomum.	*Cinnamomum.*
——— sassafras.	*Sassafras.*
Lavandula spica.	*Lavandula.*
Leontodon taraxacum.	*Dens leonis.*
Linum usitatissimum.	*Linum.*
Magnesia.	{ *Magnesia calcinata.* ——— *usta.*
Malva sylvestris.	*Malva.*
Marrubium vulgare.	*Marrubium.*
Melaleuca leucadendron.	*Cajeputa.*
Melissa officinalis.	*Melissa.*
Meloë vesicatorius.	*Cantharis.*
Mentha piperita.	*Mentha piperitis.*
——— pulegium.	*Pulegium.*
——— viridis.	*Mentha sativa.*
Menyanthes trifoliata.	*Trifolium palustre.*
Mimosæ catechu extractum.	{ *Catechu.* *Terrâ japonica.*
——— niloticæ gummi.	*Gummi arabicum.*
——— niloticæ mucilago.	*Mucilago arabici gummi.*
Mistura carbonatis calcis.	*Potio cretacea.*
Momordicæ elaterii succus spissatus.	*Elaterium.*
Murias ammoniæ.	{ *Sal ammoniacus.* *Ammonia muriata.*
——— ammoniæ et ferri.	{ *Ferrum ammoniacale.* *Flores martiales.* *Ens veneris.*
——— antimonii.	*Butyrum antimonii.*
——— barytæ.	*Barytes muriatus.*

Systematic names.	Ancient names.
Murias hydrargyri.	*Hydrargyrus muriatus corrosivus.* *Mercurius corrosivus sublimatus.*
——— sodæ.	*Muria.* *Sal marinus.*
Myristicæ moschatæ nucleus fructûs.	*Nux moschata.*
Myroxyli peruiferi balsamum.	*Balsamum peruvianum.*
Myrtus pimenta.	*Pimento.* *Piper jamaicense.*
Nicotiana tabacum.	*Nicotiana.*
Nitras argenti.	*Causticum lunare.*
——- potassæ.	*Nitrum.* *Sal nitri.*
Olea europæa.	*Oliva.*
—-- volatilia.	*Olea essentialia.*
Oleum ammoniatum.	*Linimentum volatile.*
——- succini purissimum.	*Oleum succini rectificatum.*
——— sulphuratum.	*Balsamum sulphuris.*
——— volatile pini laricis.	*Oleum terebinthinæ.*
——- ——— —- purissimum.	—— ————- *rectificatum.*
——- ——— juniperi communis.	—— *essentiale juniperi.*
Origanum majorana.	*Majorana.*
Ovis arietis adeps.	*Sevum ovillum.*
Oxalis acetosella.	*Acetosella.*
Oxidum antimonii cum phosphate calcis.	*Pulvis antimonialis.*
——— antimonii cum sulphure per nitratem potassæ.	*Crocus antimonii.* —— *metallorum.*

Systematic names.	Ancient names.
Oxidum antimonii cum sulphure vitrificatum.	*Vitrum antimonii.*
——— antimonii vitrificatum cum cera.	——— ——— *ceratum.*
——— arsenici.	*Arsenicum.* / ——— *album.*
——— ferri nigrum.	*Ferri squamæ.*
——— ——— rubrum.	*Colcothar vitrioli.* / *Crocus martis.*
——— hydrargyri cinereum.	*Mercurius calcinatus.*
——— ——— rubrum per acidum nitricum.	——— *præcipitatus ruber.*
——— plumbi album.	*Cerussa.*
——— ——— rubrum.	*Minium.*
——— ——— semivitreum.	*Lithargyrus.* / ——— *auri.*
——— zinci.	*Flores zinci.*
——— ——— impurum.	*Tutia.*
——— ——— ——— præparatum.	——— *præparata.*
Phosphas calcis impurus.	*Cornu cervi ustum.*
Physeteris macrocephali sevum.	*Spermaceti.*
Pilulæ aloës et colocynthidis.	*Pilulæ cochiæ.*
——— aloës et myrrhæ.	——— *rufi.*
——— aloëticæ.	——— *saponaceæ.*
——— ammoniareti cupri.	——— *cupri.*
——— assæ fœtidæ compositæ.	——— *fœtidæ.*
——— hydrargyri.	——— *mercuriales.*
——— opiatæ.	——— *thebaicæ.*
——— rhei compositæ.	——— *stomachicæ.*
Pimpinella anisum.	*Anisum.*

Systematic names.	Ancient names.
Pini abietis resina sponte concreta.	*Pix burgundica.*
—— balsameæ resina liquida.	*Balsamum canadense.*
—— laricis resina liquida.	*Terebinthina veneta.*
—— laricis oleum volatile.	*Oleum terebinthinæ.*
—— sylvestris resina empyreumatica.	*Pix liquida.*
Pistacia lentiscus.	*Mastiche.*
Plumbum.	*Saturnus.*
Polygala senega.	*Seneka.*
Polygonum bistorta.	*Bistorta.*
Polypodium filix mas.	*Filix mas.*
Potassa.	*Alkali causticum* *Causticum commune acerrimum.* *Kali purum.* *Lapis infernalis.* —— *septicus.*
——— cum calce.	*Causticum commune mitius.*
Prunus domestica.	*Prunus gallica.*
Pterocarpi santalini lignum.	*Santalum rubrum.*
———— dracontis resina.	*Sanguis draconis.*
Pulvis asari compositus.	*Pulvis asarabaccæ.*
----- carbonatis calcis compositus.	—— *cretaceus.*
----- cinnamomi compositus.	---- *aromaticus.* *Species aromaticæ.*
----- ipecacuanhæ et opii.	*Pulvis doveri.*
----- super-sulphatis aluminæ et potassæ compositus.	—— *stypticus helvetii.*
Quassia simaruba.	*Simarouba.*
——— excelsa.	*Quassia.*
Quercus robur.	*Quercus*

Systematic names.	Ancient names.
Resina pini.	*Resina alba.*
Rhamnus catharticus.	*Spina cervina.*
Rheum palmatum.	*Rhabarbarum.* / *Rheum.*
Rhododendron crysanthum	*Rhododendron.*
Rhus toxicodeudron.	*Toxicodendron.*
Ricinus communis.	*Ricinus.*
Rosa canina.	*Cynosbatus.*
Rosa damascena.	*Rosa centifolia.* / *—— pallida.*
---- gallica.	*—— rubra.*
Rosmarinus officinalis.	*Rosmarinus.*
Rubia tinctorum.	*Rubia.*
Ruta graveolens.	*Ruta.*
Salix fragilis.	*Salix.*
Salvia officinalis.	*Salvia.*
Sambucus nigra.	*Sambucus.*
Sapo.	*Sapo albus hispanus.*
Scilla maritima.	*Scilla.*
Sinapis alba.	*Sinapi album.*
Smilax sarsaparilla.	*Sarsaparilla.*
Solanum dulcamara.	*Dulcamara.*
Solutio muriatis barytæ.	*Solutio terræ ponderosæ salitæ.*
——— sulphatis cupri composita.	*Aqua styptica.*
Spigelia marilandica.	*Spigelia.*
Spiritus ætheris nitrosi.	*Spiritus nitri dulcis.*
——— cari carui.	*——— carvi.*
——— lauri cinnamomi.	*Aqua cinnamomi spirituosa.* / *Spiritus cinnamomi.*
——— myristicæ moschatæ.	*——— nucis moschatæ.*
——— myrti pimentæ.	*——— pimentæ.*

Systematic names.	Ancient names.
Spongia officinalis.	*Spongia.*
Styracis benzoin balsamum.	*Benzoïnum.*
Styrax officinalis.	*Styrax.*
Sub-acetis cupri.	*Ærugo.*
Sub-boras sodæ.	*Borax.*
Sub-murias hydrargyri	*Calomelanos.* *Calomelas.* *Hydrargyrus muriatus mitis.* *Mercurius dulcis.*
——— ——— et ammoniæ.	*——— præcipitatus albus.* *Calx hydrargyri alba.*
Sub-sulphas hydrargyri flavus.	*Mercurius emeticus flavus.* *Turpethum minerale.*
Succus cochleariæ officinalis compositus.	*Succi ad scorbuticos.*
——— spissatus conii maculati.	*Extractum cicutæ.*
——— ——— momordicæ elaterii.	*Elaterium.*
——— ——— sambuci nigræ.	*Rob sambuci.*
Sulphas aluminæ exsiccatus.	*Alumen ustum.*
——— barytæ.	*Terra ponderosa vitriolata.*
——— cupri.	*Cuprum vitriolatum.* *Vitriolum cæruleum.* *——— romanum.*
——— ferri.	*Sal martis.* *Vitriolum viride.*
——— magnesiæ.	*Magnesia vitriolata.* *Sal catharticus amarus.* *— epsomiensis.*
——— potassæ.	*Tartarum vitriolatum.*
——— potassæ cum sulphure.	*Sal polychrestus.*
——— sodæ.	*— glauberi.*

Systematic names.	Ancient names.
Sulphas zinci.	*Sal vitrioli.* *Vitriolum album.* *Zincum vitriolatum.*
Sulphur sublimatum.	*Sulphuris flores.*
Sulphuretum antimonii.	*Antimonium.* *Stibium.*
——— ——— præparatum.	*Antimonium præparatum.*
——— ——— præcipitatum.	*Sulphur antimonii præcipitatum vel auratum.*
——— hydrargyri nigrum.	*Æthiops mineralis.*
——— hydrargyri rubrum.	*Cinnabaris factitia.*
——— potassæ.	*Hepar sulphuris.*
Super-sulphas aluminæ et potassæ.	*Alumen.*
Super-sulphas aluminæ et potassæ exsiccatus.	——— *ustum.*
Super-tartris potassæ.	*Cremor tartari.* *Crystalli tartari.* *Tartarus purificatus.*
Super-tartris potassæ impurus.	——— *crudus.*
Suis scrofæ adeps.	*Axungia porcina.*
Syrupus althææ officinalis.	*Syrupus ex althæa.*
——— amomi zingiberis.	——— *zingiberis.*
——— citri aurantii.	——— *corticis aurantii.*
——— citri medicæ.	——— *limonum.*
——— papaveris somniferi.	——— *papaveris albi.*
——— rhamni cathartici.	——— *spinæ cervinæ.*
——— rosæ damascenæ.	——— *rosæ solutivus.*
——— toluiferæ balsami.	——— *balsami tolutani.*

Systematic names.	Ancient names.
Tamarindus indica.	*Tamarindus.*
Tanacetum vulgare.	*Tanacetum.*
Tartris antimonii.	*Tartarus emeticus.* / *Antimonium tartarizatum.*
——— potassæ.	*Tartarum solubile.*
——— ——— et sodæ.	*Sal rupellensis.* / *Sel de seignette.*
Tinctura aloës et myrrhæ.	*Elixir proprietatis.*
——— amomi repentis.	*Tinctura cardamomi.*
——— aristolochiæ serpentariæ.	——— *serpentariæ.*
——— ferulæ assæ fœtidæ.	——— *fœtida.*
——— benzoin composita	*Balsamum traumaticum.*
——— camphoræ.	*Elixir camphoræ.* / *Spiritus vinosus camphoratus.*
——— cinchonæ officinalis.	*Tinctura corticis peruviani.*
——— ——— composita.	——— ——— ——— *composita.* / ——— *alexipharmica.*
——— convolvuli jalapæ.	——— *jalapæ.*
——— eleutheriæ.	——— *cascarillæ.*
——— gentianæ composita.	——— *amara.* / *Elixir stomachicum.*
——— veratri albi.	*Tinctura hellebori albi.*
——— guajaci ammoniata.	——— *guaiaci volatilis.*
——— lauri cinnamomi.	——— *cinnamomi.*
——— lauri cinnamomi composita.	——— *aromatica.*
——— meloës vesicatorii.	——— *cantharidum.*
——— mimosæ catechu.	——— *japonica.*
——— muriatis ammoniæ et ferri.	——— *martis.*
——— muriatis ferri.	——— ——— *salita.*

Systematic names.	Ancient names.
Tinctura opii.	*Tinctura thebaica.*
	Laudanum liquidum.
—— opii camphorata.	*Elixir paregoricum.*
	—— asthmaticum.
—— rhei et aloës.	*—— sacrum.*
—— --- et gentianæ.	*Tinctura rhei amara.*
—— --- palmati.	*—— rhei.*
—— saponis.	*Balsamum saponaceum.*
	Linimentum saponaceum.
—— saponis et opii	*Balsamum anodynum.*
—— sennæ composita.	*Elixir salutis.*
—— toluiferæ balsami.	*Tinctura balsami tolutani.*
—— veratri albi.	*—— hellebori albi.*
Toluiferæ balsami balsamum.	*Balsamum tolutanum.*
Tormentilla erecta.	*Tormentilla.*
Triticum æstivum.	*Triticum.*
Trochisci carbonatis calcis.	*Trochisci cretæ.*
—— glycirrhizæ cum opio.	*—— glycirrhizæ compositi.*
Tussilago farfara.	*Tussilago.*
Unguentum acetitis plumbi.	*Uuguentum saturninum.*
—— acidi nitrosi.	*—— oxygenatum.*
—— hydrargyri.	*—— mercuriale fortius.*
	—— cæruleum.
—— —— mitius	*—— mercuriale mitius.*
—— infusi meloës vesicatorii.	*—— epispasticum mitius.*
—— nitratis hydrargyri.	*—— citrinum.*
—— oxidi plumbi albi.	*—— album.*
	—— è cerussa.
—— —— zinci.	*Unguentum zinci.*
—— —— —— impuri.	*—— tutiæ.*

Systematic names.	Ancient names.
Unguentum pulveris meloës vesicatorii.	*Unguentum epispasticum ortius.*
———— resinosum.	———— *basilici flavi.*
———— sub-acetitis cupri.	———— *æruginis.*
———— sub-muriatis hydrargyri et ammoniæ.	———— *calcis hydrargyri albi.* ———— *è mercurio præcipitato.*
Valeriana officinalis.	*Valeriana sylvestris.*
Veratrum album.	*Helleborus albus.*
Vinum aloës socotorinæ.	*Tinctura sacra.*
——— ferri.	*Vinum chalybeatum.*
——— gentianæ compositum.	——— *amarum.*
——— ipecacuanhæ.	*Tinctura ipecacuanhæ.*
——— tartritis antimonii.	*Vinum antimoniale.* *Essentia antimonialis.*
Viola odorata.	*Viola.*
Vitis viniferi fructus siccatus.	*Uva passa.*

LATIN INDEX.

....

A.

B.

C.

D.

E.

F.

G.

H.

I.

J.

K.

L.

M.

N.

O.

Q.

R.

S.

T.

U.

V.

X.

Z.

ENGLISH INDEX.

A.

B.

C.

D.

E.

F.

G.

H.

K.

L.

M.

N.

P.

Z

V.

W.

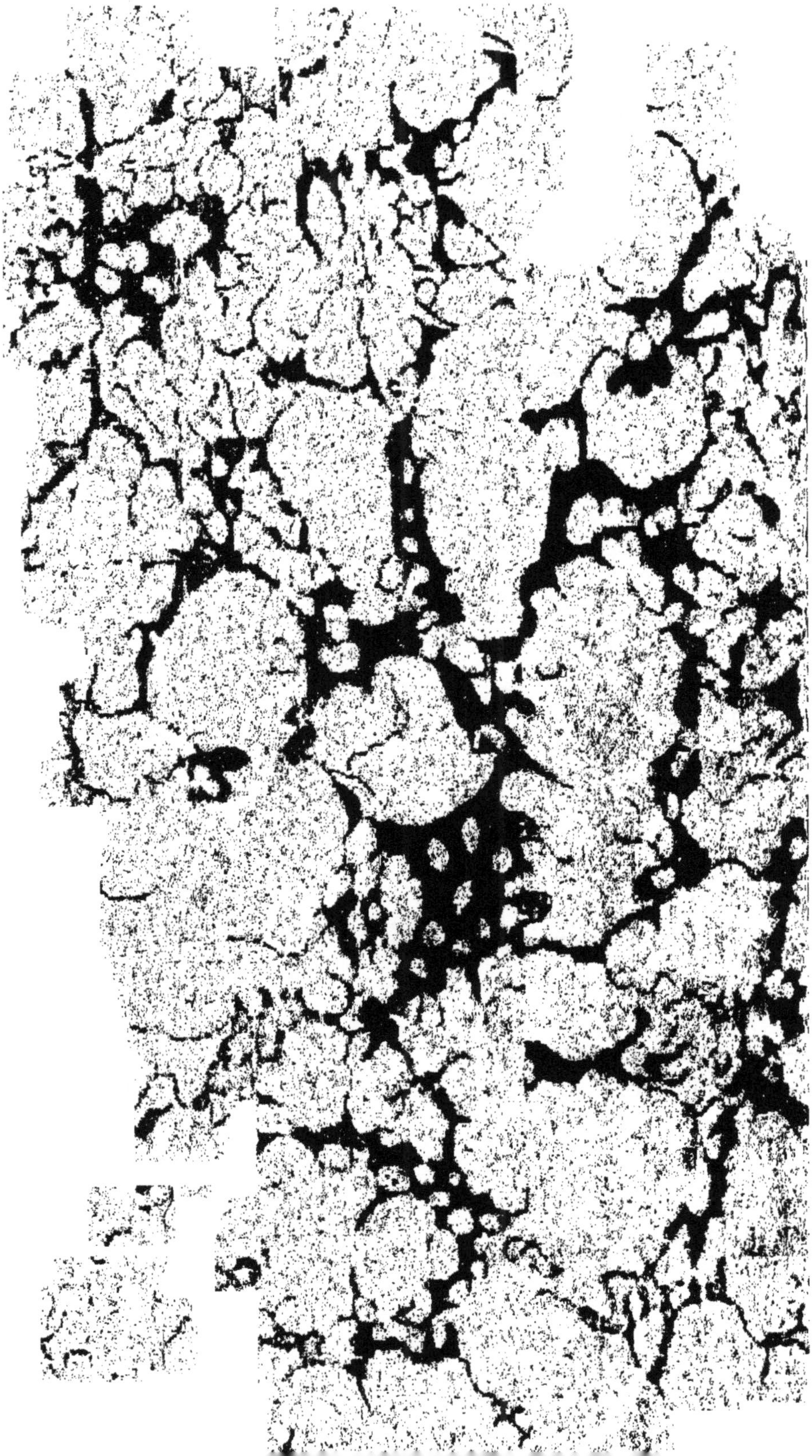

Lightning Source UK Ltd.
Milton Keynes UK
UKHW021512090219
336936UK00007B/1030/P

9 780266 822950